GOVERNMENT PROCEDURES AND OPERATIONS

SPECTRUM MANAGEMENT AND POLICY IN THE U.S.

EMERGING ISSUES

GOVERNMENT PROCEDURES AND OPERATIONS

Additional books in this series can be found on Nova's website under the Series tab.

Additional e-books in this series can be found on Nova's website under the e-book tab.

GOVERNMENT PROCEDURES AND OPERATIONS

SPECTRUM MANAGEMENT AND POLICY IN THE U.S.

EMERGING ISSUES

WANDA FLORES
EDITOR

nova publishers
New York

Library of Congress Cataloging-in-Publication Data

ISBN: 978-1-63482-265-7

Published by Nova Science Publishers, Inc. † New York

CONTENTS

PREFACE

This book discusses emerging issues in the spectrum management and policies in the United States.

Chapter 1 - The Middle Class Tax Relief and Job Creation Act of 2012 (P.L. 112-96, signed February 22, 2012) contained provisions in Title VI that expedite the availability of spectrum for commercial mobile broadband. The provisions in Title VI—also known as the Public Safety and Spectrum Act, or the Spectrum Act—cover reallocation of spectrum, new assignments of spectrum rights, and changes in procedures for repurposing spectrum used by the federal government. The act established a process for television broadcasters to release spectrum licensed to them for auction as commercial licenses (Broadcast Incentive Auctions). The act also included provisions to apply future spectrum license auction revenues toward deficit reduction; to establish a planning and governance structure to deploy public safety broadband networks, using some auction proceeds for that purpose; and to assign additional spectrum resources for public safety communications.

Broadband capacity to support popular mobile services and devices, such as real-time viewing of video on smartphones, can be improved in several ways. Examples include (1) providing new spectrum for networks to expand; (2) investing in denser infrastructure; (3) developing new technologies; or (4) expanding opportunities for sharing spectrum. Provisions of the Spectrum Act focus on increasing the amount of spectrum as the key policy tool for spectrum management.

Going forward, other policy tools may need to be considered in order to make spectrum access more inclusive. Many policy makers and Members of Congress are concerned, for example, that the current structure of auctions to assign spectrum licenses does not provide enough opportunities for competition or new entrants into mobile communications markets. These concerns include the availability of spectrum for uses such as telemedicine or driverless vehicles. The spectrum needs of emerging technologies that some believe may be key drivers of future economic growth are not specifically addressed in the Spectrum Act and appear to receive scant attention from policy makers. In addition to autonomous vehicles, growth industries that are, at least in part, spectrum dependent include advanced robotics, cloud computing, and machine-tomachine communications (the Internet of Things).

Chapter 2 - The convergence of Internet and mobile technologies has, within the last decade, transformed wireless communications and created a dynamo of innovation and economic growth. The list of applications that enable products and provide services through smart wireless devices is long and growing rapidly as new industries incorporate wireless technologies into their products. Wireless and mobile telecommunications products include not only smart phones and tablets but also utility meters, road traffic sensors, robots, autonomous vehicles, unmanned aerial systems, tractors, and household appliances, to cite but a few examples of existing and new technologies that are widely predicted to bring about profound changes in how Americans work and play. The composition of the wireless telecommunications industry is changing as companies with important stakes in spectrum-dependent technologies move from innovation to implementation.

The arrival of these technologies is accompanied by a crowd of policy questions covering issues such as employment, training, education, privacy, cybersecurity, and research and development. This report focuses on the interaction between technological change and spectrum policy, and how the accelerating pace of change may require a timely transition to new spectrum policies. Emerging technologies may require, or work better with, new network concepts to carry wireless transmissions over distances long or short. The arrival of new products, new services, and new concepts in network design may lead to the introduction of new models of competition and investment that might benefit from new spectrum policies. Spectrum policy today focuses on the expansion of commercial broadband with the goal of continuing recent growth trends attributed to the mobile Internet. Expanding policy to more fully include other technologies—including those being developed for the Fifth Generation (5G) of wireless communications—might advance a new telecommunications environment with greater potential for spurring innovation, competition, and economic growth than what has been observed in recent years.

This report traces the current and possible future evolution of mobile communications networks and some of the changes in spectrum policy that might better accommodate innovation. Congress at present is engaged in debates over how to maximize the value—economic, monetary, or other—of upcoming auctions for spectrum licenses, notably the Broadcast Incentive Auction required by the Spectrum Act in 2012 (P.L. 112-96, Title VI). The evolution of wireless technologies, as outlined in this report, indicates that auctions, as presently structured, are a limited policy tool. Congress, therefore, may move to reconsider the current goals of spectrum policy to more fully accommodate the development of the next generation of wireless technologies. In future reviews of communications law and spectrum policy, Congress may choose to broaden its scope to include spectrum-dependent industries and technologies beyond the telecommunications sector.

Chapter 3 - This is an edited, reformatted and augmented version of a white paper issued April 1, 2014.

As the first step toward modernizing the laws governing the communications and technology sector. The primary body of law regulating this sector was passed in 1934 and has not been substantially updated in 18 years. Changes in technology and the rate at which they are occurring warrant an examination of whether, and how, communications law can be rationalized to address the 21st century communications landscape. For this reason, the committee initiated an examination of the regulation of the communications industry, and seeks comments from all interested parties on the future of the law.

Chapter 4 - Radio frequency spectrum is a natural resource used to provide a variety of communication services, such as mobile voice and data. The popularity of smart phones, tablets, and other wireless devices among consumers, businesses, and government users has increased the demand for spectrum. FCC takes a number of steps to promote efficient and effective use of spectrum. One such step is to establish buildout requirements, which specify that an entity granted a license must begin using the assigned spectrum within a specified amount of time or face penalties, such as loss of the license.

GAO was asked to review buildout requirements and the efficient use of spectrum. This report (1) describes the buildout requirements FCC established for wireless services, (2) assesses the extent to which FCC follows its process to enforce buildout requirements, and (3) examines stakeholder opinions on the extent that commonly cited goals for buildout requirements have been met. GAO reviewed FCC regulations and guidance on buildout requirements and examined FCC license data on outcomes of buildout requirements for 5 out of about 45 wireless services selected to ensure variety in type of use and buildout requirement, among other criteria. GAO also interviewed FCC officials, commercial spectrum licensees, industry associations, and spectrum policy experts.

GAO is making no recommendations in this report. FCC reviewed a draft of this report and provided technical comments that GAO incorporated as appropriate.

In: Spectrum Management and Policy in the U.S. ISBN: 978-1-63482-265-7
Editor: Wanda Flores

Chapter 1

SPECTRUM POLICY: PROVISIONS IN THE 2012 SPECTRUM ACT*

Linda K. Moore

SUMMARY

The Middle Class Tax Relief and Job Creation Act of 2012 (P.L. 112-96, signed February 22, 2012) contained provisions in Title VI that expedite the availability of spectrum for commercial mobile broadband. The provisions in Title VI—also known as the Public Safety and Spectrum Act, or the Spectrum Act—cover reallocation of spectrum, new assignments of spectrum rights, and changes in procedures for repurposing spectrum used by the federal government. The act established a process for television broadcasters to release spectrum licensed to them for auction as commercial licenses (Broadcast Incentive Auctions). The act also included provisions to apply future spectrum license auction revenues toward deficit reduction; to establish a planning and governance structure to deploy public safety broadband networks, using some auction proceeds for that purpose; and to assign additional spectrum resources for public safety communications.

Broadband capacity to support popular mobile services and devices, such as real-time viewing of video on smartphones, can be improved in several ways. Examples include (1) providing new spectrum for networks

* This is an edited, reformatted and augmented version of a Congressional Research Service publication R43256, prepared for Members and Committees of Congress, dated October 1, 2014.

to expand; (2) investing in denser infrastructure; (3) developing new technologies; or (4) expanding opportunities for sharing spectrum. Provisions of the Spectrum Act focus on increasing the amount of spectrum as the key policy tool for spectrum management.

Going forward, other policy tools may need to be considered in order to make spectrum access more inclusive. Many policy makers and Members of Congress are concerned, for example, that the current structure of auctions to assign spectrum licenses does not provide enough opportunities for competition or new entrants into mobile communications markets. These concerns include the availability of spectrum for uses such as telemedicine or driverless vehicles. The spectrum needs of emerging technologies that some believe may be key drivers of future economic growth are not specifically addressed in the Spectrum Act and appear to receive scant attention from policy makers. In addition to autonomous vehicles, growth industries that are, at least in part, spectrum dependent include advanced robotics, cloud computing, and machine-tomachine communications (the Internet of Things).

WHAT IS SPECTRUM?

Electromagnetic spectrum, commonly referred to as radio frequency spectrum or wireless spectrum, refers to the properties in air that transmit electric signals and, with applied technology, can deliver voice, text, and video communications. The allocation and assignment[1] of radio frequency spectrum are managed by the Federal Communications Commission (FCC) for commercial and other nonfederal uses and by the National Telecommunications and Information Administration (NTIA) for federal government use.

Wireless broadband,[2] with its rich array of services and content, requires new spectrum capacity to accommodate growth. Spectrum capacity is necessary to deliver high speed, high quality communications to consumers and businesses and also to support the communications needs of industries that use spectrum-dependent technologies.

Although radio frequency spectrum (air) is abundant, usable spectrum is currently limited by the constraints of applied technology. Spectrum policy therefore requires making decisions about how radio frequencies will be allocated and who will have access to them. Current spectrum policy is based on managing channels of radio frequencies to avoid interference.[3] The FCC, over many years, has developed and refined a system of exclusive licenses for users of specific frequencies. Auctions are a market-driven solution to assigning licenses to use specific frequencies and are a recent innovation in spectrum management and policy. Previously, the FCC granted licenses using a process known as "comparative hearings" (also known as "beauty contests"), and has used lotteries to distribute spectrum licenses. The FCC also allocates spectrum for designated purposes, such as WiFi, without assigning a license to a specific owner (unlicensed spectrum).

For purposes of allocation and assignment, spectrum is segmented into bands of radio frequencies measured in cycles per second, or hertz. Standard abbreviations for measuring frequencies include kHz—kilohertz or thousands of hertz; MHz—megahertz, or millions of hertz; and GHz— gigahertz, or billions of hertz. The designation can refer to an entire band, such as the 700 MHz band, or to specific frequencies within a band. Most licenses are issued on a geographic basis, serving a specific area (license coverage). More than one license, therefore, is likely to have the same frequency designation.

Technology for commercial mobile services generally is designed to operate most efficiently on frequencies below 3 GHz, although the ceiling has moved higher as technology has changed, so that frequencies through 5 GHz are now considered desirable for commercial network deployment. Similarly, the bandwidth of assigned frequencies is increasing. With new technology, wider bandwidths have been shown to reduce interference and mitigate other problems hampering the efficient use of spectrum.

Spectrum Policy Provisions in the Middle Class Tax Relief and Job Creation Act of 2012

Provisions of Title VI of the Middle Class Tax Relief and Job Creation Act of 2012 (P.L. 112-96, signed into law on February 22, 2012) included expediting auctions of licenses for spectrum designated for mobile broadband; authorizing incentive auctions, which would permit television broadcasters to receive compensation for steps they might take to release some of their airwaves for mobile broadband; requiring that specified federal holdings be auctioned or reassigned for commercial use; and providing for the availability of spectrum for unlicensed use. Many of the provisions in Title VI, frequently referred to as the Public Safety and Spectrum Act, or Spectrum Act, focus on spectrum assignment within the existing regulatory framework, in which licenses for designated radio frequencies are awarded through competitive bidding systems (auctions).

Major provisions in the Spectrum Act that are summarized in this report cover

- Deficit reduction;
- Directed auctions;
- Incentive auctions for television broadcasters;
- Reallocation of spectrum from federal to commercial use; and
- Unlicensed spectrum.

Other provisions in the act, not covered in this report, include simplifying the approval of zoning requests for modification of cell towers at the state and local level[4] and putting in place measures to facilitate antenna placement on federal property. The act also has provided for the establishment of a new authority to plan and develop a nationwide public safety broadband network (FirstNet) and has included other measures in support of improved emergency communications.[5]

Deficit Reduction

The Spectrum Act has addressed the interlaced issues of spectrum access and deficit reduction. The issues are connected because, when radio frequency spectrum licenses are auctioned for commercial purposes by the FCC, the net proceeds are deposited in the U.S. Treasury.[6] The act has extended the FCC's auction authority until the end of FY2022. Because the FCC's authority would have expired at the end of FY2012, revenue from auctions held after FY2012 is considered new revenue.

The legislation that first authorized the FCC to establish "competitive bidding systems"[7] for a limited period was included in the Omnibus Budget Reconciliation Act of 1993 (P.L. 103-66). The Balanced Budget Act of 1997 gave the FCC auction authority until September 30, 2007. This authority was extended to September 30, 2011, by the Deficit Reduction Act of 2005 and to 2012 by the DTV Delay Act (P.L. 111-4). The Deficit Reduction Act of 2005 also specified that $7.363 billion of proceeds from auctions required by the act be applied to deficit reduction.

Distribution of Proceeds from Auctions Required by the Spectrum Act

Most of the proceeds from auctions of licenses in designated spectrum as specified in the act are to be deposited directly into a Public Safety Trust Fund, created by the act, with nearly $28 billion designated for purposes defined in the act,[8] including $20.4 billion for deficit reduction.[9]

Proceeds from the sale of licenses of repurposed federal spectrum identified in the Spectrum Act will be directed first to the Spectrum Relocation Fund, to cover costs of moving federal users, with the balance going to the Public Safety Trust Fund. Proceeds from the sale of advanced wireless service licenses in the other spectrum bands identified by the act will go directly to the Public Safety Trust Fund. Proceeds from the auction of new licenses created by the release of television broadcasting spectrum will go to cover costs specified in the act, with the balance to the Public Safety Trust Fund.[10] Balances remaining in any fund created by the act will revert to the Treasury in 2022.[11]

The Public Safety Trust Fund

The law provides for transfers from a Public Safety Trust Fund that is created by the act to receive revenues from designated auctions of spectrum licenses. A major beneficiary of funding is FirstNet (First Responder Network Authority), the nationwide broadband network to be developed for public safety communications.

Auction proceeds are to be distributed in the following priority:

- To the NTIA, to reimburse the Treasury for funds advanced to cover the initial costs of establishing FirstNet: not to exceed $2 billion.
- To the State and Local Implementation Fund for a grant program, as part of FirstNet: $135 million.
- To the Network Construction Fund for FirstNet, for costs associated with building the nationwide network and for grants to states that qualify to build their own networks: $7 billion, reduced by the amount advanced to establish FirstNet in the first round of funding.
- To the National Institute of Standards and Technology (NIST) for public safety research: $100 million.
- To the Treasury for deficit reduction: $20.4 billion.
- To the NTIA and the National Highway Traffic Safety Administration for a grant program to improve 911 services: $115 million.
- To NIST for public safety research, phase two: $200 million.

- To the Treasury for deficit reduction: any remaining amounts from designated auction revenues.

Directed Auctions

The Spectrum Act required the FCC and the NTIA to identify specific bands for auction from spectrum designated for commercial advanced wireless services and for federal use, and in most cases to commence the auction process within three years. The act mandated spectrum license auctions for frequencies at 1915-1920 MHz; 1995-2000 MHz; 2155-2180 MHz; an additional 15 MHz to be identified by the FCC; and 15 MHz of spectrum between 1675 and 1710 MHz, subject to conditions in the act. The Secretary of Commerce was required to submit a report to the President identifying 15 MHz of spectrum between 1675 and 1710 MHz for reallocation from federal to nonfederal use.[12] The NTIA reaffirmed an initial recommendation to reassign 1695- 1710 MHz and submitted a report, as required by the act, recommending that the FCC reallocate the band for commercial use.[13]

The act required that these auctions be completed and licenses issued by February 22, 2015, which would require that the auctions commence no later than the third quarter of 2014.[14] These licenses would provide at least 65 MHz of new spectrum for commercial broadband.

The FCC scheduled the first designated auction for January 2014. Frequencies at 1915-1920 MHz and 1995-2000 MHz, known as the H Block, were offered in Auction 96. The reserve price was set at $1.56 billion, that is, the combined final bids for all licenses offered must total at least $1.56 billion.[15] The licenses cover 176 Economic Areas.[16] Auction 96 was completed on February 27, 2014, with a total winning bid of $1.564 billion.[17]

The FCC is preparing to auction spectrum licenses for an additional 65 MHz of frequencies (Auction 97) that would complete the mandate for certain auctions by 2015.[18] The frequencies cover 1695-1710 MHz; 1755-1780 MHz; and 2155-2180 MHz. The proposed reserve price (minimum acceptable bid value) is $10.6 billion, of which $5.1 billion is to be applied to the costs of relocation or sharing of frequencies now used by the federal government.

Broadcast Incentive Auctions

The Spectrum Act has permitted the FCC to conduct incentive auctions, that is, to establish a mechanism whereby spectrum capacity may be relinquished for auction by some license-holders, who would then share in the proceeds.[19] Many commercial wireless licenses can be resold directly by their license-holders for comparable uses; the purpose of incentive auctions is to reward license-holders, such as television broadcasters, who repurpose their spectrum for a different use. Although incentive auctions might be used for other types of license-holders, the act specifically addresses spectrum assignments for over-the-air television broadcasters.

The act established procedures and guidelines for the FCC to follow in reallocating television broadcasting spectrum licenses for commercial auction. Through a reverse auction process, the broadcasters would establish the amount of compensation they are willing to accept for the spectrum they voluntarily release for auction. Additionally, broadcasters that do not voluntarily relinquish spectrum rights, but are required to relocate or incur certain other costs, may be compensated. In lieu of cash payments as compensation for relocation, broadcasters may choose to accept regulatory relief that would allow new uses for their spectrum.

Spectrum voluntarily released by TV broadcasters would be repurposed for commercial broadband communications, with licenses sold through what the law refers to as a "forward auction." At least one successful reverse auction is required to set minimum prices for a forward auction. For the results of a forward auction to be valid, auction proceeds must at a minimum cover (1) payments to broadcasters that relinquished spectrum for auction, (2) the costs to the FCC of conducting the auctions, and (3) the estimated costs for relocation of other broadcasters, which are not to exceed $1,750 million, deposited in a TV Broadcaster Relocation Fund for relocation costs. If auction revenues do not cover costs as specified in the act, the FCC may not assign new licenses and planned reassignments and reallocations may not occur. If the reverse auction and forward auction conditions are met, the FCC may "make such reassignments of televisions channels" as appropriate in its consideration, subject to certain conditions. Examples of conditions include a general prohibition against reassigning licenses to frequencies from one band to a band below an existing assignment, and obligations to determine that a reassigned channel is not adversely affected by cross-border channel assignment agreements with Canada and Mexico. The auction and channel reassignment process may only occur once.

Among the many challenges the FCC faces in establishing incentive auctions is identifying a band plan that will meet the many goals and requirements of the legislation. The band plan must identify which portion of the broadcaster spectrum should be released and how it will be divided into licenses that can be successfully auctioned.

To increase participation in the Broadcast Incentive Auction,[20] actions the FCC announced on May 15, 2014,[21] include

- Creating smaller license coverage areas within EAs (called Partial Economic Areas, PEAs);
- Assigning licenses for 5 MHz of paired spectrum, with an initial term of 12 years;
- Applying various measurements of spectrum ownership intended to handicap the ability of Verizon and AT&T to acquire licenses in certain areas.
- Reconsidering qualifications for Designated Entities based on a separate proceeding that may subsequently be applied to the Broadcast Incentive Auction;
- Requiring interoperability across the entire 600 MHz band;
- Applying to the 600 MHz band the same technical rules as for the 700 MHz band.

The outcome of the forward auction for spectrum licenses depends on the results of the reverse auction in which the broadcasters agree to the price at which they will release spectrum. Based on the amount of spectrum released by the broadcasters, a block of spectrum will be set aside in each market with competitive bidding open to Sprint, T-Mobile, and any entrant deemed not to have national network coverage. If, for example, 60 MHz of spectrum is made available by broadcasters, 20 MHz will be reserved; as described by the staff report, all registered bidders would be eligible to compete for licenses in the remaining 40 MHz. Verizon and AT&T may also be allowed to bid on reserved licenses, if there is insufficient bidding activity.

To increase innovation and usage of unlicensed spectrum, the FCC also announced steps intended to maximize the amount of spectrum available for TV White Spaces.

Federal Spectrum Use and Reallocation

The Spectrum Act addressed how spectrum resources might be repurposed from federal to commercial use through auction or sharing, and how the cost of such reassignment would be defined and compensated, among other provisions.[22] The Commercial Spectrum Enhancement Act of 2004 (P.L. 108-494, Title II) was amended to facilitate the transfer of spectrum rights to commercial purchasers from the agencies relinquishing spectrum. (See following section.) Expenditures incurred by federal agencies for planning may now be included among those costs eligible for reimbursement as part of the transfer of spectrum to the commercial sector. Other reimbursable costs cover a wide range of technical options, including spectrum sharing. Although spectrum sharing to facilitate the transition from federal to commercial use is supported in the act's provisions, the NTIA has been required to give priority to reallocation options that assign spectrum for exclusive, nonfederal uses through competitive bidding.

The act required the establishment of a Technical Panel within the NTIA to review transition plans that each federal agency must prepare in accordance with provisions in the act. The Technical Panel is required to have three members qualified as a radio engineer or technical expert. The Director of the Office of Management and Budget, the Assistant Secretary of Commerce for Communications and Information Administrator of the NTIA), and the Chairman of the FCC have been required to appoint one member each. A full discussion and interpretation of provisions of the act as regards the technical panel and related procedural requirements such as dispute resolution have been published by the NTIA as part of the rulemaking process.[23]

Commercial Spectrum Enhancement Act of 2004

The Commercial Spectrum Enhancement Act of 2004 put in place statutory rules for covering the costs to federal agencies of relocating wireless communications facilities to new spectrum assignments. The act created the Spectrum Relocation Fund to provide a means for federal agencies to recover relocation costs directly from auction proceeds when they are required to vacate spectrum slated for auction. In effect, successful commercial bidders cover the costs of relocation. Among key provisions of the act were requirements that the auctions must recoup at least 110% of the costs projected by the NTIA, and that unused funds would revert to the Treasury after eight years. These provisions remain in effect. Specific frequencies were designated for immediate auction[24] by the Commercial Spectrum Enhancement Act but the law was written to apply to any federally used frequencies scheduled for reallocation and possible auction.[25]

NTIA Plans to Make Federal Spectrum Available for Commercial Use

The NTIA, with input from the Policy and Plans Steering Group (PPSG),[26] has produced a 10-year plan and timetable that identifies bands of spectrum that might be available for commercial wireless broadband service. As part of its planning efforts, the NTIA prepared a "Fast Track Evaluation" of spectrum that might be made available in the near future.[27] Specific recommendations were to make available 100 MHz of spectrum within bands 3550-3650 MHz. The FCC has opened a proceeding for the 3.4GHz band, which may be turned into an "innovation band" for small cell networks.[28] The fast track evaluation also recommended studying two 20 MHz bands to be identified within 4200-4400 MHz for possible repurposing.

Working through the PPSG, the NTIA studied federal spectrum use by more than 20 agencies with over 3,100 separate frequency assignments in the 1755-1850 MHz band.[29] After evaluating the multiple steps involved in transferring current uses and users to other frequency locations, the NTIA concluded that it would cost $18 billion to clear federal users from all 95 MHz of the band. Based on this assessment, the NTIA report included recommendations for seeking ways for federal and commercial users to share many of the frequencies, although some frequencies were identified to be cleared for auction to the private sector. DOD estimated that relocation of its users from the 1755-1850 MHz band to clear spectrum would cost $12 billion if the bands were fully cleared. Industry estimates have valued the licenses at as much as $12 billion.

The NTIA assumptions for the estimates of the cost of relocating federal agencies from the 1755- 1850 MHz band were challenged at a hearing of the House Committee on Energy and Commerce, Subcommittee on Communications and Technology,[30] leading to a request to the Government Accountability Office (GAO) to examine the process. In particular, the NTIA was criticized during the hearing by some committee members for not separately evaluating the 1755-1780 MHz band, which might be auctioned separately with another spectrum band already available for commercial use. At the hearing, the GAO provided testimony regarding its preliminary findings on spectrum sharing[31] and followed up with a report.[32] Both the hearing and the report indicated that spectrum sharing technology and policies were largely undeveloped. Some of the options to encourage sharing spectrum, as identified by the GAO, include considering spectrum usage fees to provide economic incentive for more efficient use and sharing; identifying more spectrum that could be made available for unlicensed use; encouraging research and development of technologies that can better enable sharing; and improving and expediting regulatory processes related to sharing. Given the challenges for implementing spectrum sharing policies, the GAO found that further study by the NTIA and the FCC was needed.

GAO Cost Estimates for Spectrum Reallocation

In a hearing before the Senate Committee on Armed Services, Subcommittee on Strategic Forces,[33] the GAO presented preliminary findings on DOD estimates of reallocation costs from some radio frequencies.[34] The GAO evaluated DOD relocation cost estimates of $12 billion for frequencies at 1755-1850 MHz and reported that the "preliminary cost estimate substantially or partially met GAO's identified best practices." In particular, the GAO noted the variable nature of a number of assumptions for costs and revenues, such as the characteristics of the spectrum to which services would be relocated, the availability of new technology, and market demand for spectrum.

Release of Spectrum for Auction

Federal frequencies at 1695-1710 MHz and 1755-1780 MHz will be reassigned for commercial use on a shared basis with federal incumbents in Auction 97.[35] The frequencies at 1755-1780 MHz are to be paired with unencumbered spectrum already allocated for commercial use at 2155- 2180 MHz; auction of the latter frequencies is required by the Spectrum Act. The released federal frequencies are to be shared indefinitely with multiple federal incumbents, although some systems are to migrate to other frequencies. The frequencies at 1695-1710 MHz will be auctioned as unpaired spectrum on a shared basis with weather satellites systems. Licensees are obligated to coordinate with federal users to avoid harmful interference and to meet other requirements of transition plans.[36] The transition plans are posted on the NTIA website.[37]

Unlicensed Spectrum

Unlicensed spectrum is not sold to the highest bidder and used for the services provided by the license-holder but is instead accessible to anyone using wireless equipment certified by the FCC for those frequencies. Both commercial and noncommercial entities use unlicensed spectrum to meet a wide variety of monitoring and communications needs. Suppliers of wireless devices must meet requirements for certification to operate on frequency bands designated for unlicensed use. Examples of unlicensed use include garage door openers and WiFi communications. WiFi provides wireless Internet access for personal computers and handheld devices and is also used by businesses to link computer-based communications within a local area. Links are connected to a high-speed landline either at a business location or through hotspots. Hotspots are typically located in homes or convenient public locations.

New technologies are being developed by companies in various industry sectors to expand the usefulness of unlicensed spectrum without causing interference. For example, to use unassigned spectrum, known as white spaces, between broadcasting signals of digital television, geolocation database technology is being put in place to identify unencumbered airwaves. Similar technologies will be used to expand the availability of spectrum for unlicensed use at 5 GHz. Commercial providers, such as for wireless Internet, currently share parts of the spectrum at 5 GHz with federal users. With the objective of improving future WiFi capacity, the Spectrum Act required new studies and evaluations of frequencies at 5 GHz.[38] The FCC was required to commence a proceeding that might open access for some unlicensed devices in the 5350-5470 MHz band.[39] The NTIA was required to prepare an evaluation of spectrum-sharing technologies for the 5350-5470 MHz and 5850-5925 MHz bands.[40] Pursuant to the findings of the two agencies, the FCC has issued a *Report and Order*[41] that adds 100 MHz of spectrum for unlicensed use by changes in usage at frequencies ranging from 5.15-5.85 GHz part of the band. Future actions are likely to release additional segments of the band.

MEASURES TO IMPROVE COMPETITION

Implementation of the Spectrum Act raises a number of policy issues such as how to structure incentive auctions and how best to utilize spectrum assigned for federal use. Inherent to these debates is the question of how to provide access to spectrum on a competitive basis. Actions taken by the FCC that are intended to improve competitive access to spectrum include modifying FCC auction rules to provide licenses with smaller area coverage; encouraging participation by designated entities; and establishing limitations on the number of licenses available to Verizon or AT&T. Spectrum caps might limit the amount of spectrum available through auction to the top two carriers in key auctions such as those for repurposed television spectrum.

Geographic Coverage of Spectrum Licenses

A number of comments and petitions for reconsideration have been filed to request that the FCC include licenses for Cellular Market Areas (CMAs) in its auction rules, arguing that smaller licenses are more affordable to small, primarily rural, carriers and more nearly match the location of their target customer base. For example, using auction rules that allowed only for Economic Areas (EAs), the FCC created 176 licenses for the H Block auction (Auction 96), as compared to a potential 700 CMA licenses. For the Broadcast Incentive Auction, the FCC proposes to create Partial Economic Area licenses within Economic Areas, to allow for a greater choice between urban and rural coverage among bidders for licenses.[42]

Spectrum Caps and Screens

The history of spectrum caps as a policy to promote competition dates to preparations for the first spectrum license auctions. The FCC decided to set caps on the amount of spectrum any one company could control in any geographically designated market. The FCC's support of spectrum capping was based on the theory that each license has an economic value and a foreclosure value. The economic value is derived from the return on investment in spectrum licenses and network infrastructure. The foreclosure value is the value to a wireless company that already has substantial market share and wants to keep its dominant position by precluding competition. Spectrum caps were chosen as the method to prevent foreclosure bidding. The intent was to ensure multiple competitors in each market and to restrict bidding to only the licenses that could be used in the near term.

Beginning in 2001, spectrum policy placed increased emphasis on promoting spectrum and market efficiency. The FCC ruled to end spectrum caps, citing greater spectral efficiency from larger networks as one benefit of the ruling. Spectrum caps were seen as barriers to mergers within the wireless industry, to the growth of existing wireless companies, and to the benefits of scale economies. The spectrum caps were eliminated on January 1, 2003.[43] Auction rules requiring the timely build-out of networks became a key policy tool to deter hoarding. The FCC instituted a policy for evaluating spectrum holdings on a market-by-market, case-by-case basis— a practice referred to as spectrum screening—as a measure of competitiveness.

Current proposals to reinstate spectrum caps reflect different perspectives on competition and the role of spectrum license auctions. Although some would like to see Verizon and AT&T excluded from participation in the Broadcast Incentive Auction, the Spectrum Act prohibits exclusion of any qualified bidder.[44] How this approach might be implemented and how limitations might affect auction revenue have been the topic of spirited debate. Numerous position papers have argued effectively that any form of spectrum cap would depress auction revenues or, from a different perspective, that spectrum caps would have no impact on revenue and might even enhance it.[45] The Department of Justice has filed comments regarding the important role of spectrum access for competition and innovation. It has urged the FCC to promulgate auction rules that would prevent bids based on the foreclosure value of spectrum and that would enhance bidding opportunities for smaller carriers.[46]

A separate FCC Report and Order will establish new guidelines on the amount of spectrum that any one carrier may hold.[47] The FCC plans to apply its new criteria to the incentive auction by placing restrictions on bidding activity intended to handicap the ability of Verizon and AT&T to acquire licenses in certain areas. The majority of the 700 MHz band commercial licenses were purchased at auction in 2008 (Auction 73) by Verizon and AT&T,[48] which together also hold approximately 70% of commercial spectrum licenses below 1000 MHz.[49] The other two national carriers, Sprint (majority-controlled by SoftBank, Corp., a Japanese telecommunications provider) and T-Mobile, Inc. (majority-owned by Deutsche Telekom, AG), own 15% of commercial licenses below 1000 MHz. Although Sprint and T-Mobile are generally viewed as the primary beneficiaries of the bidding rules, the rules are also intended to provide opportunities for smaller carriers to bid successfully. Based on the amount of spectrum released by the broadcasters, a block of spectrum will be set aside in each market with bidding priority for Sprint, T-Mobile, and any entrant deemed not to have national network coverage. If, for example, 60 MHz of spectrum is made available by broadcasters, 20 MHz will be reserved; as described by the staff report, all registered bidders would be eligible to compete for licenses not acquired in priority bidding as well as in the remaining 40 MHz.

Designated Entities

One tool that the FCC regularly uses to improve competition among current and potential wireless network providers is the establishment of bidding credits for smaller companies, referred to as Designated Entities. A Designated Entity meets established criteria for size and revenue and is awarded a credit against the purchase price of an auctioned license, based on these criteria.[50] Presently, small businesses with average gross revenue of no more than $40 million in the preceding three years receive a credit of 15%. A very small business, with revenue of $15 million or less over three years, receives a credit of 25%. Rules governing eligibility for designated entity status include restrictions on the use of spectrum assets acquired through a successful bid at auction. For example, currently not more than 25% of spectrum assets acquired with a bidding credit can be leased to a larger partner with which the company has what is described as an attributable material relationship (AMR).

The FCC will reconsider the rules for designated entities in an upcoming proposed rulemaking proceeding.[51] Among the rules to be reevaluated are those for AMR, which the FCC believes may be discriminating against entrepreneurs that have formed partnerships with larger companies. In preparation for the Broadcast Incentive Auction, the FCC has granted a waiver to a private equity company that exceeds the income limits; this may set a precedent for future waivers that will widen the pool of potential bidders through more generous interpretations of designated entity rules.[52] The history of Designated Entity status and benefits—especially for minority ownership—of relaxing the rules are detailed in a report by the Minority Media & Telecom Council (MMTC).[53]

Planning for Future Needs

The Spectrum Act employs three key policy tools for increasing the availability of radio frequency spectrum for wireless broadband: allocating additional spectrum; reassigning spectrum to new users; and opening up spectrum for unlicensed use. Other policy options that may be employed to increase spectrum capacity include requiring that wireless network infrastructure be shared; changing the cost structure of spectrum access; moving to more spectrum-efficient technologies; and sharing spectrum.

Facilitating the adoption of new wireless technologies that enable spectrum sharing is emerging as a major policy consideration for spectrum management.

Policies to auction licenses for exclusive use of spectrum appear to have generally prevailed in spectrum management planning. However, a number of new spectrum-dependent industries are emerging that do not seem well served by current policies. These industries include advanced robotics, autonomous vehicles, cloud computing, and machine-to-machine communications (the Internet of Things). Spectrum resources are also needed, for example, for smart grid communications, railroad safety,[54] water conservation,[55] the safe maintenance of critical infrastructure industries,[56] and for many other applications that may not have an immediate commercial value but can provide long-lasting value to society as a whole.

Without abandoning competitive auctions, spectrum policy may benefit from including additional ways to assign or manage spectrum that might better serve the deployment of a broader range of wireless technology and the implementation of national policies.[57] Congress may consider these and other policy options as it evaluates how to meet future spectrum needs.

End Notes

[1] Spectrum is allocated for a type of use, such as television broadcasting or advanced wireless services, and assigned as licenses to specific users.

[2] Broadband refers here to the capacity of the radio frequency channel. A broadband channel can quickly transmit live video, complex graphics, and other data-rich information as well as voice and text messages, whereas a narrowband channel might be limited to handling voice, text, and some graphics.

[3] With technologies that rely on channel management, two signals can interfere with each other even if they are not at the same frequency, but are close in frequency. To avoid harmful interference, the signals must have frequencies that are sufficiently different, known as a "minimum separation."

[4] Discussed in CRS Report RS20783, Broadband Deployment: Legal Issues for the Siting of Wireless Communications Facilities and Amendments to the Pole Attachment Rule, by Kathleen Ann Ruane.

[5] Measures in the act that apply to public safety are covered in CRS Report R42543, The First Responder Network (FirstNet) and Next-Generation Communications for Public Safety: Issues for Congress, by Linda K. Moore.

[6] 47 U.S.C. §308 (j) (8). Net proceeds are the auction revenues minus the FCC's expenses. Congress has twice in the past amended the provision in order to use auction proceeds for other purposes by creating special funds to hold and disburse auction proceeds. The Commercial Spectrum Enhancement Act, Title II of P.L. 108-494, created the Spectrum Relocation Fund; the Deficit Reduction Act of 2005 created the Public Safety and Digital Television Transition Fund.

[7] 47 U.S.C. §308 (j) (3).
[8] P.L. 112-96, §6413.
[9] P.L. 112-96, §6413, 126 Stat. 235-236.
[10] P.L. 112-96, §6401 126 Stat. 223-224.
[11] P.L. 112-96, §6413, 126 Stat. 236.
[12] P.L. 112-96, §6401, 126 Stat. 223.
[13] Department of Commerce, Identification of 15 Megahertz of Spectrum Between 1675 and 1710 MHz for Reallocation from Federal Use to Non-Federal Use Pursuant to Section 640 (a) of the Middle Class Tax Relief and Job Creation Act of 2012; Report to the President, February 2013, at http://www.ntia.doc.gov/report/2013/report-presidentidentification-15-mhz-spectrum-between-1675-mhz-and-1710-mhz-reallocati.
[14] See FCC, "Remarks of Commissioner Jessica Rosenworcel, CTIA 2013—The Mobile Marketplace, Las Vegas, Nevada, May 22, 2013, at http://www.fcc.gov/document/commissioner-rosenworcels-speech-ctia-2013.
[15] FCC Public Notice, "Auction of H Block Licenses in the 1915-1920 MHz and 1995-2000 MHz Bands," AU Docket No. 13-178, September 13, 2013, http://www.fcc.gov/article/da-13-1885a1.
[16] An Economic Area (EA) is a geographic area established by the Bureau of Economic Analysis of the Department of Commerce and used by the FCC to define the coverage of spectrum licenses for certain services. There are 172 EAs, plus three EA-like areas, encompassing the Northern Mariana Islands, Guam, American Samoa, the United States Virgin Islands, and Puerto Rico. FCC map at http://wireless.fcc.gov/auctions/data/maps/ea.pdf.
[17] FCC Public Notice, "Winning Bidder Announced for Auction 96," DA 14-279, February 28, 2014, http://transition.fcc.gov/Daily_Releases/Daily_Business/2014/db0228/DA-14-279A1.pdf.
[18] FCC, Auction of Advanced Wireless Services (AWS-3) Licenses Scheduled for November 13, 2014, Docket No. 14- 78, July 23, 2014, http://transition.fcc.gov/Daily_Releases/Daily_Business/2014/db0725/DA-14-1018A1.pdf.
[19] P.L. 112-96, §6402, 126 Stat. 224.
[20] FCC Docket GN 12-268; the FCC provides information on the auction preparations at http://wireless.fcc.gov/ incentiveauctions/learn-program/
[21] FCC "Staff Summary: Incentive Auction Report and Order," May 15, 2014, http://www.fcc.gov/document/staffsummary-incentive-auction-report-and-order and Report and Order, Docket No. 12-268, released June 2, 2014, http://www.fcc.gov/document/fcc-adopts-rules-first-ever-incentive-auction-0.
[22] P.L. 112-96, §6701, 126 Stat. 245 et seq.
[23] NTIA, Notice of Proposed Rulemaking, July 17, 2012, and replies, docket no. 110627357-2209-03 at http://www.ntia.doc.gov/federal-register-notice/2012/technical-panel-and-dispute-resolution-board-nprm. Final Rule, January 25, 2013, at http://www.ntia.doc.gov/federal-register-notice/2013/spectrum-relocation-final-rule-technicalpanel-and-dispute-resolution-b.
[24] Following the procedures required by the act, the FCC scheduled an auction for Advanced Wireless Services (AWS), designated Auction 66, which was completed on September 18, 2006. The AWS auction attracted nearly $13.9 billion in completed bids. The cost to move federal agencies to new spectrum locations was set at almost $936 million.
[25] The creation of the Spectrum Relocation Fund is discussed in CRS Report RS21508, Spectrum Management and Special Funds, by Linda K. Moore.

[26] Created in response to Department of Commerce recommendations to improve spectrum efficiency through better management, see http://www.ntia.doc.gov/legacy/reports/specpolini/factsheetspecpolini_06242004.htm.

[27] NTIA, An Assessment of Near-Term Viability of Accommodating Wireless Broadband Systems in the 1675-1710 MHZ, 1755-1780 MHz, 3500-3650 MHz, and 4200-4220 MHz, 4380-4400 MHZ Bands (President's Spectrum Plan Report), November 15, 2010, at http://www.ntia.doc.gov/report/2010/assessment-near-term-viability-accommodating-wireless-broadband-systems-1675-1710-mhz-17.

[28] FCC, Further Notice of Proposed Rulemaking, GN Docket No. 12-354, released April 23, 2014.

[29] U.S. Department of Commerce, An Assessment of the Viability of Accommodating Wireless Broadband in the 1755- 1850 MHz Band, March 2012, at http://www.ntia.doc.gov/report/2012/assessment-viability-accommodating-wirelessbroadband-1755-1850-mhz-band.

[30] Hearing, House of Representatives, Committee on Energy and Commerce, Subcommittee on Communications and Technology, "Creating Opportunities Through Improved Government Spectrum Efficiency," September 13, 2012.

[31] GAO, Spectrum Management: Federal Government's Use of Spectrum and Preliminary Information on Spectrum Sharing, September 13, 2012, GAO-12-1018T, at http://www.gao.gov/products/GAO-12-1018T.

[32] GAO, Spectrum Management: Incentives, Opportunities, and Testing Needed to Enhance Spectrum Sharing, November 14, 2012, GAO-13-7, at http://gao.gov/products/GAO-13-7.

[33] Hearing, Senate, Committee on Armed Services, Subcommittee on Strategic Forces, "Oversight: Military Space Programs and Views on DoD Usage of the Electromagnetic Spectrum," April 24, 2013.

[34] GAO, Spectrum Management: Preliminary Findings on Federal Relocation Costs and Auction Revenues, April 24, 2013, GAO-13-563T, at http://www.gao.gov/products/GAO-13-563T. Full report, GAO-13-472 at http://www.gao.gov/ products/GAO-13-472.

[35] FCC, Public Notice, Auction of Advanced Wireless Services (AWS-3) Licenses Scheduled for November 13, 2014, Docket Np. 14-78; "Procedures," July 23, 2014, http://transition.fcc.gov/Daily_Releases/Daily_Business/2014/db0725/ DA-14-1018A1.pdf.

[36] FCC, Joint Public Notice, Coordination Procedures in the 1695-1710 MHz and 1755-1780 MHz Bands, Docket N. 13-185, July 18, 2014, http://www.fcc.gov/document/joint-public-notice-announcing-aws-3-coordination-details.

[37] Separate links for the two bands: http://www.ntia.doc.gov/other-publication/2014/transition-plans-and-transitiondata-1695-1710-mhz-band and http://www.ntia.doc.gov/other-publication/2014/transition-plans-and-transition-data1755-1780-mhz-band.

[38] P.L. 112-96, §6406.

[39] FCC, Notice of Proposed Rulemaking: 5GHz Unlicensed Spectrum (UNII), FCC13-22, released February 20, 2013, http://www.fcc.gov/document/increased-spectrum-available-unlicensed-devices-5-ghz-band.

[40] Department of Commerce, Evaluation of the 5350-5470 MHz and 5850-5925 MHz Bands Pursuant to Section 6406 (b) of the Middle Class Tax Relief and Job Creation Act of 2012, January 2013, http://www.ntia.doc.gov/files/ntia/ publications/ntia_5_ghz_report_01-25-2013.pdf.

[41] http://www.fcc.gov/document/5-ghz-u-nii-ro

[42] "Wireless Telecommunications Bureau Seeks Comment on a Proposal to License the 600 MHz Band Using 'Partial Economic Areas,'" FCC, DA 13-2351, December 11, 2013 and FCC Public Notice, WTB Provides Details About Partial Economic Areas, Docket No. 12-268,

released June 2, 2014, http://www.fcc.gov/document/wtb-providesdetails-about-partial-economic-areas.

[43] FCC News, "FCC Announces Wireless Spectrum Cap to Sunset Effective January 1, 2003," November 8, 2001. Report and Order FCC-01-328. See Docket No. 01-14, Notice of Proposed Rulemaking, released January 23, 2001, at http://hraunfoss.fcc. gov/ edocs_public/attachmatch/FCC-01-28A1.pdf.

[44] 47 U.S.C. 309 (j) (17) (A).

[45] Many of these papers have been filed with the FCC and can be accessed under the docket for Policies Regarding Mobile Spectrum Holdings, WT Docket N0. 12-269.

[46] Ex parte submission to the FCC by the Department of Justice, In the Matter of Policies Regarding Mobile Spectrum Holdings, WT Docket No. 12-269, April 11, 2013.

[47] "FCC Adopts Revised FCC Mobile Spectrum Holdings Policies," FCC News, May 15, 2014, http://www.fcc.gov/ document/fcc-adopts-revised-mobile-spectrum-holdings-policies; Mobile Spectrum Holdings Report and Order, Docket No. 12-268, released June 2, 20014, http://www.fcc.gov/document/mobile-spectrum-holdings-report-and-order; and Small Entity Compliance Guide: Policies Regarding Mobile Spectrum Holdings, August 4, 2014, https://apps.fcc.gov/edocs_public/attachmatch/DA-14-1135A1.pdf.

[48] The major auction of licenses for the 700 MHz band was Auction 73. Some information about the top ten successful bidders in Auction 73 is available at Wireless Strategy, FCC Auctions, http://www.wirelessstrategy.com/ 700auction.html.

[49] "Fact Sheet FCC Mobile Spectrum Holdings," May 2014, FCC staff report, http://www.fcc.gov/document/fact-sheet fcc-mobile-spectrum-holdings-rules.

[50] For complete description of current rules, see http://www.law.cornell.edu/cfr/text/47/1.2110.

[51] Official FCC Blog, "Empowering Small Businesses," Roger Sherman, Wireless Telecommuni cations Bureau Chief, August 1, 2014, http://www.fcc.gov/blog/empowering-small-businesses.

[52] FCC, Order, Docket No. 12-268, released July 23, 2014, http://www.fcc.gov/document/ commission-grants-grainmanagement-waiver-section-12110b3.

[53] MMTC, Digital Déjà vu: A Road Map for Promoting Minority Ownership in the Wireless Industry, S. Jenell Trigg and Jeneba Jalloh Ghatt, February 25, 2014, http://mmtconline. org/mmtc-white-paper-wireless ownership-02-24-14/

[54] The railroad industry uses wireless communications as part of their information networks to monitor activity.

[55] For example, sensors buried at the level of plant roots recognize when watering is needed and communicate this information over wireless networks.

[56] In general, critical infrastructure industries facilitate the production of critical goods and services such as safe drinking water, fuel, telecommunications, financial services, and emergency response. A discussion of key issues appears in CRS Report RL30153, Critical Infrastructures: Background, Policy, and Implementation, by John D. Moteff.

[57] These issues are discussed in CRS Report R43595, Mobile Technology and Spectrum Policy: Innovation and Competition, by Linda K. Moore.

In: Spectrum Management and Policy in the U.S. ISBN: 978-1-63482-265-7
Editor: Wanda Flores

Chapter 2

MOBILE TECHNOLOGY AND SPECTRUM POLICY: INNOVATION AND COMPETITION*

Linda K. Moore

SUMMARY

The convergence of Internet and mobile technologies has, within the last decade, transformed wireless communications and created a dynamo of innovation and economic growth. The list of applications that enable products and provide services through smart wireless devices is long and growing rapidly as new industries incorporate wireless technologies into their products. Wireless and mobile telecommunications products include not only smart phones and tablets but also utility meters, road traffic sensors, robots, autonomous vehicles, unmanned aerial systems, tractors, and household appliances, to cite but a few examples of existing and new technologies that are widely predicted to bring about profound changes in how Americans work and play. The composition of the wireless telecommunications industry is changing as companies with important stakes in spectrum-dependent technologies move from innovation to implementation.

The arrival of these technologies is accompanied by a crowd of policy questions covering issues such as employment, training, education, privacy, cybersecurity, and research and development. This report focuses on the interaction between technological change and spectrum

---

* This is an edited, reformatted and augmented version of a Congressional Research Service publication, No. R43595, dated September 30, 2014.

policy, and how the accelerating pace of change may require a timely transition to new spectrum policies. Emerging technologies may require, or work better with, new network concepts to carry wireless transmissions over distances long or short. The arrival of new products, new services, and new concepts in network design may lead to the introduction of new models of competition and investment that might benefit from new spectrum policies. Spectrum policy today focuses on the expansion of commercial broadband with the goal of continuing recent growth trends attributed to the mobile Internet. Expanding policy to more fully include other technologies—including those being developed for the Fifth Generation (5G) of wireless communications—might advance a new telecommunications environment with greater potential for spurring innovation, competition, and economic growth than what has been observed in recent years.

This report traces the current and possible future evolution of mobile communications networks and some of the changes in spectrum policy that might better accommodate innovation. Congress at present is engaged in debates over how to maximize the value—economic, monetary, or other—of upcoming auctions for spectrum licenses, notably the Broadcast Incentive Auction required by the Spectrum Act in 2012 (P.L. 112-96, Title VI). The evolution of wireless technologies, as outlined in this report, indicates that auctions, as presently structured, are a limited policy tool. Congress, therefore, may move to reconsider the current goals of spectrum policy to more fully accommodate the development of the next generation of wireless technologies. In future reviews of communications law and spectrum policy, Congress may choose to broaden its scope to include spectrum-dependent industries and technologies beyond the telecommunications sector.

ECONOMIC GROWTH, INNOVATION, AND DEMAND FOR RADIO FREQUENCY SPECTRUM

By most measures the United States leads the world in the deployment of mobile broadband technology[1]—networks, devices, applications, products, and services—creating a mini-boom of innovation and entrepreneurship, and associated wealth and jobs.[2] Innovations in mobile broadband use the Internet Protocol (IP) to bring new devices, applications, and technologies to the airwaves. Cellular Long Term Evolution (LTE) and Wi-Fi are key standards that provide mobile broadband access to the Internet and IP-enabled networks.

Projections of significant growth in demand for mobile broadband—and, hence, for spectrum capacity—have prompted numerous policy initiatives to identify radio frequencies that can be made available for commercial mobile broadband services. Finding additional spectrum for commercial mobile broadband has been a goal for both the Administration and Congress. The Federal Communications Commission (FCC)—which manages radio frequency spectrum for commercial and other non-federal uses—and the National Telecommunications and Information Administration (NTIA)—which represents federal government users—are addressing some regulatory and policy issues, such as those identified in the National Broadband Plan (NBP).[3] For example, the NBP cited spectrum capacity as a critical input for expansion of mobile broadband and set a goal of providing 500 MHz of additional spectrum by 2020, to support expected growth. President Barack Obama issued a Presidential Memorandum[4] in support of the NBP's spectrum goals and, among other actions, instructed the NTIA to work with the FCC in identifying spectrum used by federal agencies to transfer to the commercial sector.[5]

The 112th Congress addressed some of the issues of spectrum allocation and assignment in the Middle Class Tax Relief and Job Creation Act of 2012 (P.L. 112-96, Title VI, "Spectrum Act").[6] The primary intent of the spectrum-clearing provisions of the act appears to be to increase the number of exclusive-use spectrum licenses for mobile broadband coverage. Among its provisions, the Spectrum Act included mechanisms to reassign television broadcast spectrum (Broadcast Incentive Auction) for licenses to be auctioned, and to transfer spectrum assets from federal to commercial use.

The act also addressed the need for unlicensed spectrum, heavily used for Wi-Fi communications, among other applications. It included allowances for unlicensed spectrum in planning for repurposed broadcast spectrum, and called for the release of spectrum in the 5 GHz band[7] for unlicensed use. Many of the frequencies at 5 GHz are currently allocated for federal use.

As new wireless technologies are introduced and perfected, access to spectrum capacity is becoming essential to doing business and to life-style choices. Debates within Congress regarding spectrum policy cover a wide range of concerns, such as competition, economic growth, access to wireless services, affordability, and opportunities for new entrants. Much of the debate appears based on assumptions about wireless technology and spectrum demand and capacity that may not fully recognize the ever-accelerating pace of technological innovation.

Policy goals that focus on supporting commercial mobile broadband may favor commercial wireless carriers at the expense of other industry sectors that will require spectrum for future growth. Policy makers largely recognize the need to strike a balance between meeting immediate needs to sustain growth for mobile broadband and finding solutions to support future growth in other wireless technologies, such as those discussed below and in the **Appendix**. Nonetheless, there appears to be a significant and widening gap between what emerging wireless technologies require and current spectrum policy delivers. Faced with this apparent policy vacuum, the 113th Congress has posed the question of what changes may be required in order to provide sufficient spectrum capacity for the future.[8]

Future questions from Congress may introduce a wide-based discussion about established goals for competition and access,[9] and about the demands on spectrum capacity made by new technologies.[10] In addition to mobile broadband, many other technologies that promise substantial growth through innovation also require access to spectrum. Policy makers may choose to explore how spectrum policy can support sustainable innovation and growth in emerging wireless technologies through better management of spectrum resources. Policy makers involved in the debate on spectrum policy may wish to see Congress take a leadership role in identifying the changes in law, regulation, and policy that are needed to fuel broad-based growth of wireless technologies. Others may prefer a minimal role for Congress and greater latitude for market forces to resolve the challenges presented by new technology.

SPECTRUM-DEPENDENT INDUSTRIES

A May 2013 study by McKinsey & Company[11] identified a dozen "disruptive technologies" and the industries that are likely to significantly change the ways that people and businesses organize their work and personal life styles. At least five of the technologies identified in the report (McKinsey report) are dependent on licensed and unlicensed spectrum to be fully functional. These technologies and others offer opportunities for economic expansion but often bring with them shifts in demand for human capital, unmet needs for new skills, challenges to security and privacy, and other changes that may need to be addressed with new public policies. Similarly, new policies and regulations may be needed to make room for innovation, while equitably preserving the value of existing infrastructure, where possible.

One of the impacts of disruptive technologies is on business models: new ways of business flourish and old business practices wither. The Internet is an example of a disruptive technology. Its impact is far-reaching and still growing. One of the early impacts of the Internet was to change and eventually diminish the role of proprietary network technologies. Business plans that were built on attaining market share through control of a superior proprietary network were eroded or made obsolete. Today, some maturing companies that built business plans on expectations of how the Internet would be deployed are faced with accommodating shifts in market demand from the wired Internet to the mobile Internet.[12] Today, the IP standard developed for the wired Internet is being incorporated into mobile network standards; tomorrow, a new standard for the mobile Internet may emerge, setting off a new cycle of technological innovation and infrastructure investment.[13]

In the McKinsey report, technologies were evaluated according to projected increases in growth, globally. Based on the report, categories of disruptive technologies that offer significant opportunities for growth, and are dependent on access for spectrum to achieve much of that growth, are

- Mobile Internet: mobile broadband devices that access the Internet;
- Cloud technology: access to off-site computing and other information-based capabilities;
- Internet of Things: the interconnection of electronic and mechanical devices through the Internet, also known as the Internet of Everything;
- Advanced robotics: high-level robots that equal or exceed human abilities to perform tasks; and
- Autonomous and near-autonomous vehicles: any vehicle, land-based, airborne, or maritime, that moves with little or no human intervention.

In addition, other industry sectors cited by the McKinsey report as affected by disruptive technologies may also rely in part on wireless technology; for example, the electric power grid requires wireless communications to manage information about demand and usage; and gas and oil exploration and production use wireless sensors.

A summary of the McKinsey report discussion of selected disruptive technologies appears in the **Appendix**.

Although not all-inclusive, the list indicates the wide range of spectrum-dependent technologies and the critical role they may play in economic growth. A number of these technologies were originally developed by the Defense Advanced Research Projects Agency (DARPA).

Other than the mobile Internet, the McKinsey report did not explore the dependence of these technologies on wireless communications, or discuss their need for spectrum capacity. Broadband for the mobile Internet may be an essential part of mobile technology but, in the long run, its future development may have less impact on U.S. economic growth and productivity than mobile technologies used to support emerging technologies, such as those described, by McKinsey, as disruptive.

The benefits of innovation are rarely evenly distributed. There are often unintended consequences and some sectors of society and the economy may suffer harm. Therefore not everyone will agree that some of the disruptive technologies discussed by McKinsey and in this report should be nurtured. The focus of this report is the value of spectrum access to a broad range of industries beyond the traditional telecommunications field. Spectrum policy may inadvertently restrict that access, thereby creating a de facto industrial policy without full consideration of the value of the affected industries and their ability to innovate.

INNOVATION AND TRANSITION

Disruptive technologies, by their nature, lead to additional innovation, new markets, new business models, new forms of competition, new investment opportunities, and other changes. Innovation and entrepreneurship may be curtailed if regulatory policies favor one technology, one market, or one business model over others. Consistent regulatory policies, however, tend to be beneficial for existing businesses and investors to the extent that they provide certainty and establish parameters for doing business. There are challenges in making the transition to new policies just as there are challenges in making the transition from one technology to another. In the case of wireless technology, successive introductions of new technology in general build on past investments in infrastructure.

Current Environment

Key factors that—separately or together—may shape current policy decisions and regulations regarding spectrum access include (1) high demand for spectrum access; (2) competitive benefits from economies of scale for wireless carriers; (3) standardization; and (4) global harmonization of spectrum allocation.

Access to Radio Frequency Spectrum

Spectrum bands have typically been allocated for a type of use, such as television broadcasting or Advanced Wireless Services (AWS), and assigned through licensing. Most commercial license assignments in the United States are now done through the auction of licenses for exclusive ownership, which conveys the right to exclusive use, although license-holders may sublet access to their holdings. Some spectrum is allocated for unlicensed use, which permits access through certification of specific devices and the enforcement of regulations limiting interference among users. Unlicensed spectrum is shared by approved devices, with access determined largely by the type of device in use. Access to licensed spectrum may also be shared. One common model for sharing today is for two or more commercial license-holders to reach a contractual agreement to share. Another common model is for federal users to permit commercial access to its assigned frequencies, often on a geographic basis (access to some areas) and/or a time basis (certain times of a day or specific days). In these shared environments, access is dependent, in part, on permission by the primary user or license-holder.

Economies of Scale

The current environment for commercial mobile network expansion appears to favor continued investment by large, often global, carriers, in LTE and LTE Advanced network infrastructure. Regulatory policy assumes that wireless communications deployments benefit from economies of scale, because of required investments in technology-driven infrastructure, among other causes. For example, U.S. wireless carriers reportedly spent over $34 billion on their networks in 2013.[14] Much of this was spent on what is sometimes referred to as the macro network, that is, the typical configuration of cell towers and base stations, linked above ground by wireless microwave transmissions and below ground by fiber-optic networks. The macro network facilitates nationwide wireless coverage. The high costs and difficulties in raising capital to cover these investments may place smaller carriers at a competitive disadvantage.[15] Furthermore, the high cost of entry may discourage new competitors. Not including the price of purchasing spectrum licenses, billions of dollars are required to build new infrastructure. The sunk costs of incumbent wireless service providers therefore set a high bar for new entrants if they are to compete effectively in major markets. Furthermore, existing wireless carriers with substantial investments in infrastructure and large customer bases are generally prepared to pay more at auction than high-risk new ventures, reinforcing barriers to new entrants.

Standards

Industry standards provide numerous benefits such as improved interoperability of systems and equipment, lower manufacturing costs, and greater ease-of-use for consumers. Standards may also be established by one industry group at the expense of other groups that may not have been active participants in the standards development process. Wireless standards adopted by market leaders for mobile broadband may place smaller rivals at a competitive disadvantage, or preclude innovation and growth in other industries.[16]

Regulators and policy makers may rely on standards-setting groups to provide technical frameworks for regulations and policy goals. Standards may be reproduced in rules and regulations—such as rules established by the FCC for spectrum license auctions—that some stake-holders view as perpetuating existing business models to the possible disadvantage of new entrants with different business models.[17]

In the United States, the two sets of standards most used for new wireless broadband technologies are cellular LTE, for licensed spectrum, and IEEE (Institute of Electrical and Electronics Engineers) 802.11, for unlicensed spectrum. There are additional IEEE standards for Wide Area Networks. An LTE standard has been developed to operate on unlicensed spectrum (LTE-U).

The main standards-setting body for LTE/LTE Advanced is the Third Generation Partnership Project (3GPP).[18] Standards organizations represented through members of 3GPP are the European Telecommunications Standards Institute (ETSI),[19] the Alliance for Telecommunications Industry Standards (ATIS),[20] and several Asian standards groups.[21]

Policy decisions about unlicensed spectrum are generally centered on providing additional capacity for channels used for Wi-Fi. Standards for Wi-Fi are developed by the IEEE 802.11 Working Group and supported by the Wi-Fi Alliance.[22] Activities of the Alliance, an industry association, include support of industry standards, certification of devices, and development of new product specifications. The 802.11 suite of standards includes 802.11.a, the original Wi-Fi standard for operations at 5 GHz and 802.11b, designated for Wi-Fi at 2.4 GHz. New spectrum assignments are being sought within the 5 GHz range[23] for the expansion of 802.11.ac VHT (Very High Transmission), sometimes known as Gigabit Wi-Fi.[24] Spectrum that has been made available for Wi-Fi on TV White Spaces is accessed by 802.11.af, sometimes referred to as White-Fi. Gigabit Wi-Fi, White-Fi, and other new standards use technologies that can identify locations of users to manage transmissions without causing interference. These standards, which are sometimes referred to as Fifth-Generation Wi-Fi, are important building blocks in developing new forms of sharing, such as between federal and commercial network operators.

Harmonization

Harmonization refers to the allocation of spectrum across national borders. Compatible designations for using spectrum facilitate economies of scale in designing and producing wireless devices. Harmonization also facilitates cross-border travel and trade, among other benefits. The global champion for harmonization is the International Telecommunications Union (ITU), the lead United Nations agency for information and communications technologies.[25] ETSI, 3GPP, and the GSM Association[26] are among the groups that work closely with the ITU in developing standards to harmonize spectrum for mobile broadband. Agreements on international harmonization are typically negotiated as treaties under the auspices of periodic World Radio Conferences (WRC),[27] supported by the ITU. There are over 30 spectrum harmonization proposals currently under consideration by the WRC. Many of these proposals would require repurposing spectrum band allocations to accommodate mobile broadband.

Proposals for harmonization and coordination of spectrum access are also being developed by national and international groups for wireless technologies for machine-to-machine communications; robotics; maritime, terrestrial, and aerial unmanned vehicles; cloud computing; and other emerging technologies.[28]

One aspect of harmonization that may be an issue is that requirements for standards may be part of the agreement. For mobile technologies, there appears to be an inclination to make LTE the default standard for harmonized frequencies. Also, harmonization of unlicensed spectrum has proved difficult to negotiate in international discussions. The global trend, therefore, might be described as harmonization for licensed spectrum assigned exclusively to network operators using LTE/LTE Advanced standards.

Transition

As wireless carriers increase their deployment of the mobile Internet, delivered by broadband LTE technologies, new trends are developing in network design and investment strategies. The development of and constant improvements in small cell technologies[29] have enabled the deployment of Heterogeneous Networks, or HetNets. In an LTE environment, a HetNet is viewed as an extension of LTE network concepts and standards. HetNets combine macro and micro infrastructure to increase the reach of wireless transmissions. The macro network is a high-site cellular network, operating on dedicated spectrum, and relying on fixed infrastructure such as towers and masts. The micro network is composed of multiple placements of low-site small cells.[30] These micro networks operate on licensed or unlicensed spectrum. Wireless carriers are investing in sites for small cell installations in order to increase network capacity, and building out micro networks to supplement or possibly replace investment in more capital-intensive macro networks. Within the LTE/LTE Advanced technology development envelope, many industry experts predict a shift in investment from building macro networks to building micro networks.[31]

Figure 1 depicts a built-out Third-Generation (3G) macro network. Its primary physical components are antennas for transmitting and receiving wireless signals—placed on towers, tall buildings, or other structures—and base stations for connecting to other communications networks, such as the Public Switched Telephone Network and the Internet. Investment in towers and base stations is critical to providing coverage and ubiquitous service to geographic areas that, typically, correspond to spectrum licenses for exclusive use.

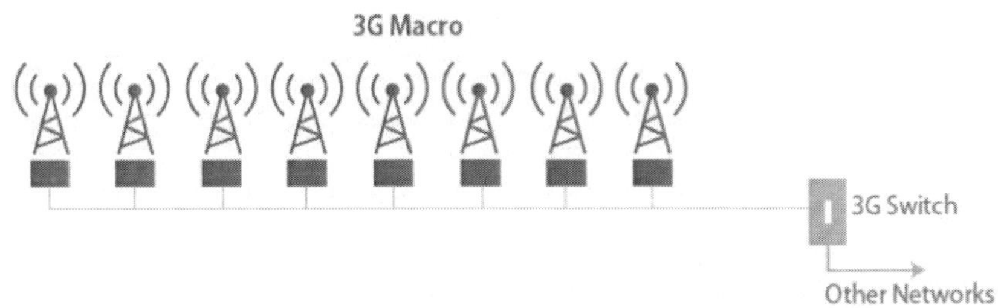

Source: Peter Rysavy research for 4G Americas, *Mobile Broadband Explosion; the 3GPP Wireless Evolution*, August 2013, Figure 57, p. 117, http://www.rysavy.com /Articles/2013-08-4G-Americas-Mobile-Broadband-Explosion.pdf.

Figure 1. 3G-Macro Network.

Figure 2 shows an example of a HetNet composed of a partly-built LTE network bolstered by a micro network of small cells being used as a bridge between 3G infrastructure and LTE towers. The micro network, for the most part, connects to the infrastructure that supports the macro network base stations.

In this simplified schematic, the 3G macro layer switch connects to the Internet and other networks through the IP gateway serving the 4G macro and micro layers. The 4G macro base stations and small cell locations are all IP-enabled.

The geographic layer represents the communities served by the HetNet. A HetNet build-out typically uses spectrum licensed to the carrier to carry traffic over both macro towers and small cell placements. Additional capacity is provided by Wi-Fi. The LTE micro network serves many of the same customers in the geographic layer as Wide Area Networks (WAN) using IEEE 802.11 (Wi-Fi/WAN).

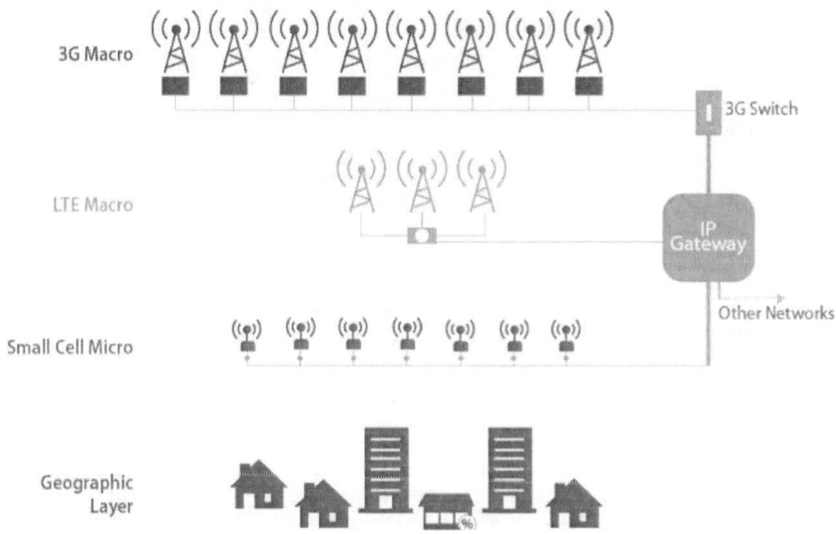

Source: Peter Rysavy research for 4G Americas, *Mobile Broadband Explosion; the 3GPP Wireless Evolution*, August 2013, Figure 57, p. 117, http://www.rysavy.com /Articles/2013-08-4G-Americas-Mobile-Broadband-Explosion.pdf.

Figure 2. LTE Heterogeneous Network.

This depiction of a HetNet configuration assumes that LTE will become the predominate technology for micro networks, using both licensed and unlicensed spectrum. The effectiveness of the HetNet in linking macro and micro network infrastructure depends in part on the proximity of IP-enabled network nodes.

The key elements of the network architecture are

- LTE macro network, which would typically be considered the transport network;
- Network interfaces such as the IP gateways;
- Micro network, which increases capacity and coverage; and
- End users and their devices, residing in the geographic layer.

The macro network provides coverage and the micro network provides capacity as well as extra coverage. In urban areas, a HetNet may carry as much as 80% of its traffic over unlicensed Wi-Fi. In the macro network, the geographic layer corresponds roughly to license coverage as designated by the FCC. Micro networks operate within geographic areas that are smaller in coverage than the licenses assigned for 3G and 4G build-outs.

Innovation

In the future, micro networks may use spectrum licensed for their purpose, and the coverage areas for these licenses may correspond to small cell neighborhoods. A possible trend in the evolution of mobile networks is depicted in **Figure 3**. In this configuration, the micro network has become the predominate provider of mobile communications coverage and capacity through contiguous small cell networks. The towers of the macro network enhance connectivity and provide additional coverage. Micro networks for small cell neighborhoods exist within or are independent of HetNets. Small cell networks are by and large autonomous, providing coverage for their area and connecting to other small cell networks or cellular towers when needed. Such a configuration might lead to widespread spectrum sharing.

The transition from macro to small cell networks as the main providers of wireless capacity would change the dynamic of spectrum demand. Low-frequency, wide coverage, high-site cellular networks that require exclusive licenses for efficient operation would no longer be the drivers of growth and change. A backbone of cellular towers would remain to serve the ever-growing, ever-changing micro networks that are more agile and more responsive to the needs of a wide range of wireless customers.

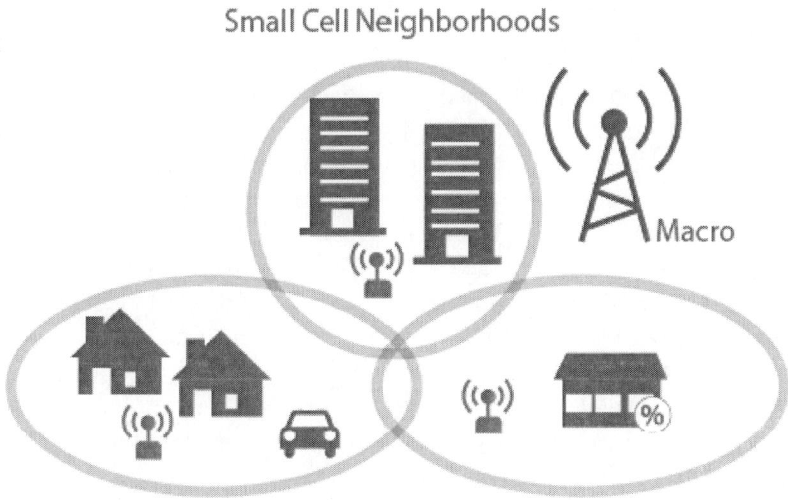

Source: 4G Americas, *Meeting the 100X Challenge: The Need for Spectrum, Technology and Policy Innovation,* October 2013, Figure AI-1, p. 130, http://www.4gamericas.org/documents/ 2013_4G%20Americas%20Meeting%20 the%201000x%20Challenge%2010%204%2013_FINAL.pdf.

Figure 3. Small Cell Neighborhood Micro Network.

As the network becomes decentralized, with more traffic carried through localized small cell networks, the towers of the macro network might be used primarily for high-value communications that require, for example, a high level of Quality of Service (QoS).[32] One attribute of QoS is the minimization of interference by using dedicated radio frequencies. Assuring QoS is one of the arguments for requiring dedicated spectrum for mobile broadband.

As the number of small cell networks multiply, increasing coverage, mobile communications will likely become less dependent on macro networks and their traffic might decline. In this case, there would likely be an over-supply of spectrum for macro networks and possibly a spectrum crunch for small cell networks. Therefore, if the majority of mobile traffic uses shared spectrum in small cell networks, the amount of dedicated spectrum required for the macro network might conceivably be reduced.

Since small cell networks are expected by most technologists to perform well in shared conditions, spectrum sharing—including, perhaps, licenses that mandate sharing—may provide more effective tools for promoting competition and growth.

Transition might occur as an evolution first from macro to micro networks and then from centralized network controls to user-controlled mobile devices, connecting through traffic management centers that will migrate to the cloud as technology permits. Today's cellular network is predicted to be one of many transport options, with small cell network physical structure integrated into devices or small cell neighborhood structures.

The combination of cloud technology and small cell networks may provide significant savings in capital expenditures for wireless companies. For example, as reported in the Financial Times,[33] AT&T is planning to move some functions of switches and routers to cloud-based, software-defined networks. Consequently, the company expects, in the future, to reduce capital expenditures on networks. The Financial Times article quoted an industry analyst who stated "AT&T is transforming the wide area network the way Information Technology transformed the data center."

The evolution of the HetNet is supported by advances in LTE and IEEE standards. Expected innovations within the LTE envelope are planned for incremental change within linear cellular network architectures. LTE/LTE Advanced, however, does not encompass the universe of wireless innovation. IEEE standards have in general been more flexible in accommodating non-cellular communications for emerging technologies, such as robotics. True innovation, it might be argued, would occur through wireless network technologies that emulate the Internet by using multiple connections that transfer traffic seamlessly from one node to another.

Disadvantages of LTE Network Architecture

Spectrum policies in general favor cellular network architecture. For example, the FCC's working definition of flexible-use service rules for spectrum is apparently based on technical rules for high-site, high power cellular antennas.[34] Although LTE standards provide the framework for high-speed, IP-based networks, LTE's limitations may be a barrier to the development of more advanced network architectures that are fully aligned with the design principles of the Internet.

A study in late 2012 by two researchers at AT&T Labs[35] suggested "a fresh new look" at design strategies and service assumptions in order to move mobile communications away from its "cellular circuit-oriented" origins. The study noted that LTE traces its heritage to the General Packet Radio Services (GPRS), designed to add packet-switched functionality to GSM, and uses the same basic approach to moving communications traffic to and from a centralized gateway. The evolution from GPRS to LTE is described as "tunnel-centric architecture." The flaws of this architecture, as described in the study, become barriers to future scaling and innovation.

The paper commented on the rise in mobile traffic and the "unexpected side effect due to interactions between cellular network architecture and the constant connectivity" required by always-on mobile devices, such as smartphones. Maintaining this connectivity strains "gateways and other network elements." To support wireless devices such as sensors (a critical element in many disruptive technologies, as discussed in the **Appendix**) using current LTE tunnel-centric architecture would require either permanent connections or the addition of "expensive" network signaling to manage usage, the researchers argued.

The Scalable Network: Innovation, Competition, Investment

Although micro networks were first seen by wireless carriers as a way to increase the capacity of their macro networks, it is becoming evident to many that micro networks can also be the primary means of providing wireless access. A small cell network can be treated as a separate business enterprise and customers within a closely inter-dependent geographic area can be targeted for additional services.[36]

To improve the effectiveness of mobile connectivity, wireless networks are themselves likely to become more mobile, flexible, and deployable (nomadic). Network infrastructure will move closer to the devices it currently supports, as the enabling technologies migrate from the macro to the small cell network. Next-generation traffic management centers will connect to corresponding nodes that connect directly to the Internet instead of through a cellular core network. To provide coverage, cell tower networks may be supplemented by satellites, or by Unmanned Aerial Systems (UAS) using, for example, drones.[37]

In the design proposal depicted in **Figure 4**, the IP-enabled core uses cloud technology to connect end users to the Internet. The widespread availability of small cell locations and multiple options for transporting traffic allows for scalability ranging, for example, from a small sports stadium to a large urban area.[38]

The key elements of this evolved network architecture are

- End users and their devices, which, using small cells and new technologies, are the main drivers of the network;
- Network interfaces operating largely through the cloud; and

- The transport layer, which now includes macro cell towers, micro small cells, and other systems (satellite, UAS) that are integrated and interchangeable.

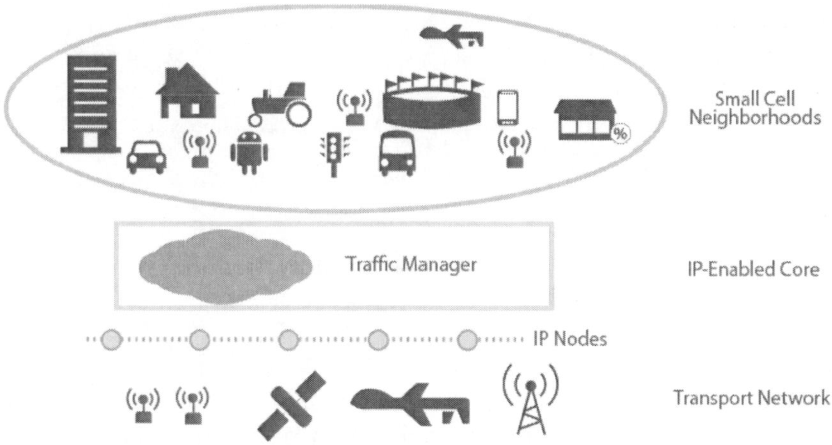

Source: Congressional Research Service (CRS).

Figure 4. Scalable Mobile Access Networks.

In the future depicted in **Figure 4**, network architecture is no longer dominated by fixed towers and cellular technology. Access is consumer-driven not network-controlled. Wireless traffic from a plethora of devices is directed through a new generation of traffic management centers to any IP-enabled transport link. The traffic management connections access most physical infrastructure through the cloud.[39] Coverage is assured through any IP-enabled transport system, whether fixed cell towers, fixed or deployable small cells, satellites, unmanned aerial systems, or other.

In this environment, economies of scale can be achieved through cloud technology, not by network consolidation and spectrum license aggregation. Ownership of exclusive-use licenses may not, therefore, be essential to assure ubiquitous coverage; may no longer be relevant for achieving important economies of scale; and may be counter to policy goals that seek to increase the capacity of spectrum for productive use. Certainty for investors may be provided through policies that balance multiple, competing interests, not through ownership of exclusive-use licenses.

By providing new opportunities for competition, changes in technology also might provide new opportunities for investment. For example, achieving economies of scale has been a driving force in the wireless industry for over a decade, rationalizing consolidation of spectrum license holdings. Micro networks, which operate in small geographic areas, may not require scale economies to be competitive. If scale is no longer critical to success, opportunities for small investors and nascent entrepreneurs may arise, spurring new types of investment models.[40] In such an environment, scale may no longer be the consequence of technological and market forces but might be viewed as an artificial business model favoring monopolistic behavior.

It is possible to envision a wireless business environment where the value for investors derives primarily from ownership of micro networks and intellectual property rights for wireless devices and applications. This shift would likely significantly change investment priorities, redirecting research and development to products that operate on micro networks and to the continuous improvement in products and services.

The Arrival of Fifth Generation Networks

Most announcements about future 5G network designs include some reference to small cell concepts, albeit at a more advanced level of technology than what is in use today. Recent descriptions of 5G emphasize important improvements in network speed and capacity and the introduction of new communications technologies. Industry comments diverge on whether 5G will remain within the suite of cellular technologies—a further advance of LTE—or represent a new direction in network architecture. Some describe 5G as not one standard but a combination of several standards and technology families or of new types of air interfaces.[41] IEEE 802.11.ac standards for 5 GHz are seen by many as an important stepping stone in the development of 5G networks that will use both licensed and unlicensed spectrum.

Commercial deployments of Fifth-Generation technologies by 2020 have been announced by wireless network officials in South Korea[42] and Japan.[43] In July 2014, Ericsson, a global leader in communications technology, demonstrated the speed of its 5G network design to customers NTT DOCOMO (Japan) and SK Telecom (South Korea).[44]

The year 2020 is the target introduction date for many companies and industry groups that have been established to advance the introduction of 5G. These include IMT-2020;[45] 5G Forum;[46] METIS;[47] the 5G Information Centre;[48] and 5G-PPP.[49]

CHANGING SPECTRUM POLICY TO ACCOMMODATE NEW TECHNOLOGY

A key component of spectrum policy is the allocation of bands for specific uses and the assignment of frequencies within those bands. The allocation and assignment of radio frequencies, particularly as it relates to the five disruptive technologies noted in this report, might today be described as operating in two distinct domains: exclusive-use commercial spectrum licenses and unlicensed spectrum. These two domains are presently dominated by two suites of standards:

- The cellular network track laid out by 3GPP where innovation occurs within the LTE envelope and relies primarily on spectrum licenses assigned for exclusive use; and
- The unlicensed track using IEEE Wi-Fi standards for channels allocated for that purpose, providing access and capacity to all comers.

The first track—a seamless network built using LTE technology—is supported by current spectrum policy, which favors auctioning spectrum licenses for exclusive use. Licenses are a bankable asset that provides some protection to investments in infrastructure, as there is reasonable certainty that investors will hold the asset long enough to realize a return.

The minimum effective bandwidth for assuring quality of service over a wide geographic area with LTE is 20 MHz, many experts say. Carrier aggregation, a feature of LTE Advanced, typically allows for operations on up to five 20-MHz bands, or 100 MHz, by as many as five carriers. Contiguous spectrum of 100 MHz in bands designated for LTE is therefore considered valuable to LTE carriers.[50] To get the full benefit of LTE Advanced coverage, macro network operators also seek licenses with large geographic areas, thereby committing to ever-greater capital investments.

The second track builds on unlicensed spectrum for Wi-Fi and related standards that are expanding to meet the needs of different industries as their technologies evolve. Access to unlicensed spectrum is an important resource for many industries—manufacturing, utilities, agriculture, and others—not generally considered part of the telecommunications industry. Unlicensed spectrum benefits from attributes of small cells, such as their ability to operate on multiple frequencies (spectrum agnostic) or to support shared use.

Under the current regulatory regime, the FCC often chooses to differentiate between macro and micro networks only in terms of licensed and unlicensed spectrum, therefore, arguably, not fully realizing or supporting the growth potential of micro networks on licensed spectrum. It appears, however, that market demand for low-cost, easy access to spectrum will grow as spectrum-dependent industries outside the telecommunications sector continue to expand. Some economists project that some of these industries are on the threshold of exponential growth.[51]

Although unlicensed technologies may be more suitable for their spectrum needs, these new industries often must choose between less suitable and often-costly LTE network solutions or spectrum access under unlicensed regulations that do not provide certainty and may discourage investment.

To meet the needs of these emerging markets and accommodate new entrants, the FCC might modify its spectrum policies, in particular its rulemaking procedures for spectrum license auctions, a key policy tool.

Spectrum Auctions

Auctions, a fairly recent innovation in frequency assignment, are regarded as a market-based mechanism for allocating spectrum. Other market-driven policies include licensing fees based on fair-market valuations of spectrum and flexibility in spectrum usage within assigned bandwidths. Today, spectrum licenses for commercial applications are typically auctioned to the highest bidder.

One ongoing spectrum policy debate centers on how to create a framework for greater competition through widespread ownership of spectrum licenses. One policy tool is the rule-making process for participation in spectrum auctions.

Provisions in existing law require the FCC to provide opportunities for small businesses to compete in auctions, which is accomplished largely by policies for Designated Entities (DE).[52] To accommodate economies of scale for macro networks, the FCC has, generally, been increasing the size of geographic coverage of licenses. To assure the licensed spectrum is put to full use, the FCC applies build-out deadlines for network infrastructure, and other requirements.

The FCC's decisions about geographical coverage for licenses are believed by some to influence the number of potential bidders. In designing band plans for spectrum to be auctioned, the FCC chooses the geographic coverage of licenses. Although there are a number of geographic configurations for licenses, the two most frequently used are designated as Economic Areas (EAs) and Cellular Market Areas (CMAs).[53] CMA coverage is based largely on Metropolitan Statistical Areas or Rural Service Areas. The Economic Area license size tends to favor business planning for larger communications networks seeking maximum geographic coverage. Because of economies of scale, their Return on Investment for spectrum license purchases is presumed to be better than the return for a local wireless service provider that may not be able to take full advantage of a geographic coverage that extends beyond its market area. Not surprisingly, larger carriers can, and often do, outbid smaller companies in acquiring licenses.[54] To counterbalance what some consider to be an unfair advantage in bidding competitions, the FCC in the past has employed spectrum caps, limiting the amount of spectrum any one carrier may acquire at auction.

Examples of the role of auctions in allocating and assigning spectrum access and rights can be found in two current FCC rulemaking proceedings. One is for the Broadcast Incentive Auction, which includes decisions on allocation of spectrum between licensed and unlicensed use, and assignment through auctions. Auction rules for the Broadcast Incentive Auction[55] favor license sizes and build-out rules for macro networks. The license coverage will be for Partial Economic Areas (PEAs) that create smaller license coverage areas within EAs, but final rules may allow bidders to aggregate licenses. The other FCC action concerns reallocation of federal spectrum holdings in the 3.5 GHz band. The FCC proposes to designate 3.5 GHz as an “Innovation Band.” In part to encourage micro networks and their successor technologies, the FCC proposes reducing the geographic coverage of a license to a single census tract. It is considering criteria for designated entities that might focus less on company size or ownership and more on programs for innovation or research and development.

Reallocation and Assignment: Broadcast Incentive Auction

The Spectrum Act has permitted the FCC to conduct incentive auctions, that is, to establish a mechanism whereby spectrum capacity may be relinquished for auction by some license-holders, who would then share in the proceeds.[56] Many commercial wireless licenses can be resold directly by their license-holders for comparable uses; the purpose of incentive auctions is to reward license-holders, such as television broadcasters, who repurpose their spectrum for a different use. Although incentive auctions might be used for other types of license-holders, the act specifically addresses spectrum assignments for over-the-air television broadcasters.

The act established procedures and guidelines for the FCC to follow in reallocating television broadcasting spectrum licenses for commercial auction. Through a reverse auction process, the broadcasters would establish the amount of compensation they are willing to accept for the spectrum they voluntarily release for auction. Additionally, broadcasters that do not voluntarily relinquish spectrum rights, but are required to relocate or incur certain other costs, may be compensated.

Spectrum voluntarily released by TV broadcasters is to be repurposed for commercial broadband communications, with licenses sold through what the law refers to as a "forward auction." At least one successful reverse auction is required to set minimum prices for a forward auction. The outcome of the forward auction for spectrum licenses depends on the results of the reverse auction. For the results of a forward auction to be valid, auction proceeds must at a minimum cover (1) payments to broadcasters that relinquished spectrum for auction, (2) the costs to the FCC of conducting the auctions, and (3) the estimated costs for relocation of other broadcasters, which are not to exceed $1.750 million, deposited in a TV Broadcaster Relocation Fund for relocation costs. Auction revenue amounts above these three financial obligations are to be deposited in a Public Safety Trust Fund created by the Spectrum Act to receive and disburse proceeds of the spectrum license auctions required by the act.

The Public Safety Trust Fund is scheduled to receive approximately $1.5 billion from Auction 96, which was completed on February 27, 2014.[57] Additional revenue will come from another auction of spectrum licenses (Auction 97) required by the act, scheduled for November 2014.[58] The proposed reserve price (minimum acceptable bid value) is $10.6 billion, of which $5.1 billion is to be applied to the costs of relocation or sharing of frequencies now used by the federal government. Mandated disbursements from the Public Safety Trust Fund include $7.135 billion for the development of a nationwide public safety broadband network, which has priority as a recipient, and $20.4 billion for deficit reduction.[59]

Participation and License Coverage

Increasing the availability of commercial broadband is the stated goal of both the Broadcast Incentive Auction and the FCC. Although the act does not specify that spectrum licenses be sold to allow for the build-out of LTE networks, LTE is the primary standard for broadband on cellular networks. Furthermore, the broadcasting frequencies that are targeted by the FCC for auction are in the 600 MHz band, a band designated for LTE and adjacent to the 700 MHz band, where LTE network build-outs have begun.

The FCC has designed rules for the Broadcast Incentive Auction[60] intended to improve competition among current and potential LTE network providers. The licenses will be for 5MHz of paired spectrum, to maximize the amount of licenses available without creating problems of interference with television broadcasts by cellular network transmissions. License coverage will be for Partial Economic Areas,[61] although the auction rules might allow for aggregation, creating even larger licensed coverage. The FCC will require interoperability for devices across all networks built in the 600 MHz band; this will likely benefit smaller carriers as well as improve the efficiency of spectrum access. Qualifications for Designated Entities will be reconsidered in a separate proceeding that may subsequently be applied to the Broadcast Incentive Auction.

A separate Report and Order will establish new guidelines on the amount of spectrum that any one carrier may hold.[62] This evaluation, which imposes limits on ownership to avoid market concentration, often referred to as a spectrum screen, is presently made on a case-by-case basis for merger activity. The FCC plans to apply its new criteria to the incentive auction by placing restrictions on bidding activity intended to handicap the ability of Verizon and AT&T to acquire licenses in certain areas. The majority of the 700 MHz band commercial licenses were purchased at auction in 2008 (Auction 73) by Verizon and AT&T,[63] which together also hold approximately 70% of commercial spectrum licenses below 1000 MHz.[64] The other two national carriers, Sprint (majority-controlled by SoftBank, Corp., a Japanese telecommunications provider) and T-Mobile, Inc. (majority-owned by Deutsche Telekom, AG), own 15% of commercial licenses below 1000 MHz and did not participate in Auction 73. Although Sprint and T-Mobile are generally viewed as the primary beneficiaries of the bidding rules, the rules are also intended to provide opportunities for smaller carriers to bid successfully. Based on the amount of spectrum released by the broadcasters, a block of spectrum will be set aside in each market with bidding priority for Sprint, T-Mobile, and any entrant deemed not to have national network coverage. If, for example, 60 MHz of spectrum is made available by broadcasters, 20 MHz will be reserved; as described by the staff report, all registered bidders would be eligible to compete for licenses not acquired in priority bidding as well as in the remaining 40 MHz.

The announced rules are consistent with past FCC auction practices over time. Testimony by FCC Chairman Tom Wheeler affirmed the FCC's assumption that economies of scale skew competition for spectrum licenses.[65] Given the predicted decline for investment in macro networks, the success of the auction would appear to depend on attracting new entrants, including entrepreneurs that are looking to serve new markets for wireless services. It is possible, however, that the technical rules developed for LTE networks in the 700 MHz band, to be applied to the 600 MHz band, may limit entrepreneurial innovation, or at least constrain it within the LTE envelope.

The auction rules appear to be giving priority to the policy goal of increasing competition among macro network transport providers, where economies of scale prevail. The FCC has determined that "low-band" spectrum is uniquely desirable for network coverage because it is "better suited for transmitting wireless communications over long distances and through walls."[66] With the new Mobile Spectrum Holdings Policies, the FCC seems to have drawn a line at 1000 MHz, below which it will encourage the build-out of LTE networks and attempt to maximize the number of exclusive-use spectrum license owners to provide network coverage. Allowances are also made to increase unlicensed use as part of the Broadcast Incentive Auction. These decisions reaffirm the two-track spectrum policy that has been in place for decades.

Reallocation and Assignment: 3.5 GHz

In its *Further Notice of Proposed Rulemaking* (FNPRM) that would repurpose spectrum at 3.5 GHz,[67] the FCC has pointed the way toward a third track of spectrum policy. The FNPRM proposes a framework for spectrum management that is considered by many to be a precursor to the development of spectrum policies more aligned with evolving wireless technologies. The proposed rules address important issues for the development of small cells and sharing technologies such as license coverage, license assignment, interoperability, competition from new entrants, and rules for the management of shared access. Many of the proposed rules are in outline form, with requests for comments specifically asking for industry guidance.

Participation and Coverage

The FNPRM proposes establishing three tiers of spectrum access in the 3550-3650 MHz band, taking advantage of advances in small cell and shared-spectrum technologies. Under the proposal, there would be a top tier of Incumbent Access for current users such as the Department of Defense for radar systems, and Fixed Satellite Service. A second tier would use a Spectrum Access System to allot spectrum among pre-qualified users, who will be assigned Priority Access (PA). The third tier would be available nationwide for General Authorized Access (GAA). Although not designated as unlicensed, the GAA layer would serve a comparable purpose. The proposal would allocate at least half of the available spectrum (after allowing for Incumbent Access) for GAA. GAA would be open to all comers, with rules established by the FCC. However, up to 20 MHz of the GAA may be reserved for hospitals, public safety organizations, local governments, or similar—designated as Contained Access Users—for indoor use within their premises. The FCC refers to tiers two and three as Citizens Broadband Radio Service.

The boundaries of the three tiers would shift depending on usage, managed, in general, through Spectrum Access Systems (SASs), that would use dynamic spectrum management technology to manage sharing throughout the band. The FCC assumes that multiple SASs would be in operation on a nationwide basis throughout the band. It proposes that SASs be allowed to collect "reasonable" fees from PA and GAA users.

The proposed rulemaking would create Priority Access Licenses (PALs) with an initial authorization for one year, for a 10 MHz assignment of a single census tract. A PAL license might be assigned directly by the FCC or won through competitive bidding. Longer license periods and license aggregation are being considered. Large players might, either at auctions or over time, be able to assemble licenses covering areas up to 30 MHz. The FCC might also, in auctioning spectrum licenses, provide a mix of license sizes. The expectation is that the availability of small cell area licenses will spur competition and innovation. The FCC is seeking comment as to whether to offer bidding credits to further encourage participation by small businesses. Still, the barrier to entry posed by legal costs may remain high for small businesses unless the FCC can further simplify both the requirements for participating in auctions and the certification of devices for license-by-rule allocations.

Economies of scale in device manufacturing are to be met by requiring interoperability of all devices for PA and GAA users across the full 100 MHz being reallocated at 3.5 GHz; requirements may be extended to an additional 50 MHz at 3650-3700 MHz that may be added later. The presumed technologies are LTE-Unlicensed and new and current forms of Wi-Fi.

As the FCC notes in the FNPRM, the effectiveness of its proposals for 3.5 GHz is largely dependent on "the development and implementation of a robust SAS." Their approach, therefore, is to propose several high-level requirements with the expectation that industry participants will develop the technical implementation and any needed standards.

The proposed rulemaking fulfills expectations as an innovation band in that it provides the opportunity to see how effectively small cell and shared technologies can increase spectrum capacity. The proposed coverage areas for PA and GAA users are limited, however, and may not be sufficient to spur competition for PALs, as the economic benefits may be elusive. At present, Incumbent Access users have pre-emptive rights to airwaves on both coasts of the United States, significantly diminishing the amount of spectrum available for licensing.

In establishing the band plan, the FCC adhered to exclusion zones originally established in 2010 by the NTIA. These zones were based on assumptions for the deployment of high-site macro networks using WiMax standards[68] as the main criteria for evaluating interference. Many argue that these criteria are meaningless in a small cell environment. Some expect that, once the three-tier designations are in place and the Spectrum Access Systems operating, the exclusion zones will gradually disappear. In the short term, however, there is concern that the desired competition and influx of new players for PA licenses may not materialize.

Whatever the perceived shortcomings of the yet-to-be-decided final rules, lessons learned from the Innovation Band can be used to develop a road map from present to future policies and practices.

EVOLUTION OF SPECTRUM POLICY

Spectrum policy, notably as interpreted by the regulations and action of the FCC and the NTIA, can accelerate or delay the arrival of the network of the future. The key factors identified in this report as influencing policy decisions appear to be changing with the technology and may lead to policy changes.[69]

Access to Radio Frequency Spectrum

The demand for access to spectrum is projected to continue to increase dramatically. The need for spectrum licenses for exclusive use to meet this demand is, however, being challenged by new technology that enables sharing. One policy proposal for sharing would permit the auction of licenses for the right to share federal spectrum. In the United States, this approach is usually referred to as Licensed Shared Access (LSA). Using dynamic spectrum access to monitor and control spectrum availability, access to cellular networks would be available either to the federal incumbent or the licensed network operator. Transitional technologies, such as those being developed for 3.5 GHz, are making possible other forms of sharing and the creation of new categories of licenses.

The question of spectrum to support 5G technologies has barely been addressed in public forums. Spectrum at 15GHz will likely be used for the 5G network scheduled for deployment in Japan in 2020. 5G introductions based on LTE standards in the United States might use existing spectrum holdings where investments have been made in LTE HetNet infrastructure.

Economies of Scale

The shift to micro networks would appear to reverse the need for large investments in infrastructure and spectrum licenses as a requisite for market participation. New forms of economies of scale might be achieved not by a few investing a great deal but by many investing comparatively small sums. If investment requirements as a barrier to entry are reduced, opportunities for new participants and competitors are likely to increase. Other barriers include lack of certainty about the availability and unit cost of spectrum access and the costs of regulatory compliance.

The importance of economies of scale in the manufacture of devices, already important, is likely to increase as devices become the drivers of the network. Interoperability across multiple bandwidths promotes scale for device manufacturers.

Standardization

Substantial resources have been invested in developing a mobile Internet that can link cellular architecture to IP-enabled networks using LTE/LTE Advanced technology and standards. New standards are also emerging for Wi-Fi, Wide Area Network, Cloud technologies, and 5G.

The trend may be shifting away from cellular network technology, although the transition is likely be gradual, given the large investment in LTE network infrastructure. Based on cellular technologies, LTE provides continuity with carriers' legacy systems. Because LTE is IP-enabled, it supports high speed broadband and can connect with most systems that operate with IP protocols.

The evolution of LTE standards is considered by many, notably policy makers within the European Union, as providing the path to future innovation. In this view, innovation occurs within the planned, predictable evolution of LTE; policies for planned growth are predicated on LTE standards for the Internet of Things, autonomous vehicles, robotic sensors, and other emerging or yet-to-be invented technologies that require access to spectrum. Standardization in a planned environment may discourage the introduction of disruptive technologies and may encourage industry cartels.

Domestic and international forums that set standards for mobile broadband may need to give greater consideration to the interdependence of standards. This would include not only the LTE and IEEE standards used in HetNets but other standards that may be crucial to expanding coverage and capacity. This would include standards for Unmanned Aerial Systems terrestrial links, and for the emerging technologies that provide for a mix of spectrum usage within a single network, such as that being tested for 3.5GHz.

Harmonization

Efforts to harmonize spectrum allocations on an international basis have and will likely continue to require reallocation of spectrum to commercial broadband, providing opportunities to further expand LTE networks. However, the diplomacy of harmonization may shift away from allocating or reallocating spectrum for LTE broadband toward placing more emphasis on harmonizing other standards, including 5G. Coordination of unlicensed spectrum and certification of devices for international use may become more important to international forums in the future.

Investment, spectrum access, and standardization interact with each other and may influence both national and global business decisions. As markets become increasingly global, access to spectrum on a worldwide basis may be one of the entry barriers for some American companies. Therefore, international negotiations regarding spectrum use are likely to play an important part in implementing wireless policies.

Internet Infrastructure

Changes in spectrum policies, especially as regards auctions, may improve access to spectrum for a broad variety of players. However, if wireless access is an important input for a growing number of industries, access to fixed communications infrastructure, notably the Internet, is crucial for mobile communications. Micro networks achieve capacity through a proliferation of small cells, each connecting to the Internet. Less space between antennas means that more connections to Internet infrastructure are required.

DEVELOPING THE SMALL CELL NETWORK

Within less than a decade wireless communications has advanced from Third Generation (3G) to Fourth (4G) and is poised to start the transition to the Fifth (5G). Although some industry experts foresee a continued build-out of LTE-enabled towers using small cells to supplement coverage, other experts foresee the evolution of interconnected but separate networks that rely primarily on small cells and emerging new technologies. Some view 5G as a continuation of the evolution of LTE macro networks; others envision 5G as a potentially disruptive micro network technology that may break entirely with its cellular origins. These alternative scenarios likely lead to different outcomes depending in part on spectrum assignment policies. Policies that favor 5G/LTE/cellular networks might imply support for the more efficient use of capital investment in existing infrastructure. Policies that favor 5G/IEEE/non-cellular solutions might be deemed to be more oriented toward innovation and new entrants.

A key goal of spectrum policy is to manage a resource to achieve the maximum benefit to society, for example by creating jobs, opportunities for investment and growth, and new forms of communications available to all.

Current policies of spectrum license assignment rely heavily on auctions, where competition to acquire licenses is judged to spur competition within the telecommunications industry and maximize revenue to the U.S. Treasury.[70] These rules largely favor telecommunications companies with major investments in macro networks.

Another policy goal is emerging: to assure spectrum access as a reliable, low cost input for dozens of industries that are not part of the telecommunications sector. This need appears to be critical for broad-based economic growth across many sectors of the economy. As described in this report, efforts to accelerate LTE/LTE Advanced have shown the advantages of using micro networks in addition to macro networks. Many technologists, as discussed throughout this report, have reasoned that the miniaturization and compression of network components potentially places more control of communications access in the hands of users (devices) and is less reliant on macro networks using exclusive-use spectrum. In a policy environment that provides certainty about spectrum availability and access, financial markets may increase investment in micro networks and the many technologies that benefit from their presence. These include the disruptive technologies identified in this report: autonomous vehicles, advanced robotics, cloud computing, and machine-to-machine communications. Other sectors and industries that benefit from reliable access to spectrum, but may not benefit from using LTE networks, include medicine, education, agriculture, retailing, utilities, and manufacturing.

A third, emerging, spectrum policy track provides a bridge between the exclusive use/everybody uses dichotomy that currently predominates. Efforts to commence the transition to a new spectrum regime are evident in the proposed FCC rules for the 3.5 GHz band. They are counter-balanced by the FCC's rulemaking for the Broadcast Incentive Auction. As described above, this rulemaking is focused on rules to accommodate macro networks. Both proceedings address issues of competition for access to spectrum primarily through regulation rather than by leveraging the opportunities provided by technology. Neither of the proceedings fully accounts for the importance of Internet access nor acknowledges the imminent transition to 5G. In the mobile world, Internet access and spectrum access are equally essential.

Whereas some of the engineering solutions that will enable the sharing of 3.5 GHz spectrum have yet to be perfected, the technologies for small cell networks are in current use. The potential exists, therefore, to increase competition and innovation with policies that recognize small cell networks as a separate and distinct sector of the telecommunications industry.

If 3.5 GHz deployments meet expectations and show a clear path forward through an evolution of small cell and spectrum sharing technologies, the United States may in the future have a surplus of exclusive-use licenses that are not being fully utilized on a competitive basis.

The potential for small cell and spectrum-sharing technologies to increase competition, and provide new opportunities for new entrants and innovation, might merit fuller consideration than what is offered through the creation of an Innovation Band. Some argue that auction rules that provide small cell area licenses and recognize different build-out and usage requirements for small cell networks would meet goals for maximum benefit and auction revenue. They are asking for a new spectrum management regime that will open access to more types of investments across a broader range of industries.[71] Since evolving 5G architectures are predicted to be heavy users of small area networks, new spectrum policies for small cells may also smooth the path to 5G.

Auctioning licenses for small cell networks might advance the third policy track and move to an environment that no longer focuses on the spectrum needs of the traditional telecommunications industry. The area coverage for these licenses should, most concur, allow for flexibility. Since the technology can support a network that provides service to, for example, a sports arena, then a license with the geographic area of a sports arena should be available, they say.

Some of the same challenges for assuring access and encouraging competition that face policy makers for spectrum also apply to the Internet. Policies to encourage consumer-driven networks may address these issues, at least in part, by creating new markets that will likely attract new infrastructure providers.

POLICY CONSIDERATIONS FOR CONGRESS

Policy challenges created by the rapid evolution of wireless technologies that Congress may decide to consider include the following:

- Developing spectrum policy goals beyond meeting immediate needs for mobile broadband.
- Identifying transitional opportunities for spectrum assignment and allocations.

- Improving the application process and reducing the cost of obtaining certification for unlicensed devices.
- Removing regulatory barriers to the development of flexible infrastructure.
- Aligning spectrum access policies with policies governing Internet access.
- Permitting technology to evolve while also balancing regulatory needs to achieve desired policy goals.

Congress may also choose to consider whether the two agencies responsible for spectrum policy—the FCC and the NTIA—are capable of fully meeting their responsibilities. Do they have the right resources to do their jobs? Are they sufficiently accountable for the consequences of their decisions? Are they adequately balancing traditional telecommunications technology with emerging and evolving technology?

APPENDIX. SPECTRUM-DEPENDENT TECHNOLOGIES

This report bases its references to disruptive technologies on a May 2013 study by McKinsey & Company.[72] The report identified a dozen "disruptive technologies" and their industries. The definitions and categorization of these technologies as provided by McKinsey serve in this report as a baseline for understanding spectrum-dependent technologies and their role in the U.S. and global economies.

Mobile Internet

Advances in wireless communications technology have fostered a large and growing market for mobile broadband services, also referred to as the mobile Internet.[73] The mobile Internet is defined by the McKinsey report as a combination of mobile computing devices, high-speed wireless connectivity, and applications. Wireless devices support voice, text, and video communications providing information and entertainment. An increasing amount of mobile communications is through social media, exchanging information among virtual communities. Mobile Internet devices and networks are enhanced by the use of the Internet Protocol for interoperable connectivity and applications development. The impact of the Internet on retailing, banking and payments, education, health services, and other public and social services is well advanced and projected by the McKinsey report to increase and expand. The development of software applications (apps) for the mobile Internet may be a separate disruptive technology.

Increasingly, as this report discusses, licensed spectrum for the mobile Internet is used for LTE cellular technology on networks owned and operated by wireless carriers. These networks cater primarily to markets for the mobile Internet but are currently expanding their customer bases in the Internet of Things, near-autonomous vehicles, and cloud technology.

Cloud Technology

Cloud technology, as described by the McKinsey report, allows the delivery of potentially all computer applications and services through networks or the Internet. Many commonly used Internet services are delivered using cloud resources, such as online searches, and streaming media. The cloud and the mobile Internet contribute to each other's growth, and cloud technology is an enabler for the Internet of Things and other wireless innovations. One of the advantages of the cloud model is elasticity. With cloud technology, notes the McKinsey report, capital-intensive investments for infrastructure can be turned into "asset light" operating costs because, for example, peak demand loads can be distributed throughout the cloud. The ability to shift demand easily throughout cloud resources provides another advantage: greater reliability. These benefits contribute to cloud technology's potential to disrupt existing business models.

The development of cloud technologies might be the source for new services and applications for mobile users. Decreased reliance on fixed infrastructure for data processing and storage might also include less dependence on fixed network infrastructure, such as the base stations that serve cellular towers.

Internet of Things

The Internet of Things, also known as machine-to machine communications, is defined in the McKinsey report as the use of sensors, actuators, and data communications technology built into physical objects, enabling the objects to be tracked, coordinated, or controlled across a data network or the Internet. The three main components for the Internet of Things are the sensors and actuators that detect and communicate information; programming and analytical software; and data communications links. Wireless sensors, including Radio Frequency Identification (RFID) chips, usually operate on designated unlicensed spectrum. Wireless connections might be by cellular networks or microwave, using licensed frequencies, or over unlicensed frequencies.

The McKinsey report includes Intelligent Transportation Systems (ITS) as part of the Internet of Things. ITS relies heavily on sensors for vehicle-to-vehicle and vehicle-to-infrastructure communications. It uses spectrum assigned to the Department of Transportation (DOT) for ITS and other programs in DOT's Research and Innovative Technology Administration (RITA). The ITS program currently focuses on development of the connected car.[74] Applications for trucks and automobiles are oriented toward traffic safety, using vehicle-to-vehicle and vehicle-toinfrastructure, short-range communications systems.

Autonomous and Near-Autonomous Vehicles

An autonomous vehicle is defined by the McKinsey report as a vehicle that can move with little or no human intervention. Some autonomous applications are already familiar, such as autopilot in airplanes and self-parking cars. Key technologies for autonomous vehicles are machine vision systems, artificial intelligence, and sensors. The signals from machine vision and sensors are integrated through artificial intelligence to provide directions to vehicles. Similar applications are used to guide robots. The images captured through machine vision can be sent over wireless communications technologies such as gigabit Ethernet. For autonomous guidance systems, Ethernet operates on unlicensed Wi-Fi frequencies or on higher frequencies for fixed (point-to-point) connections.

Unmanned Aerial Vehicles (UAV or drones) are categorized by the McKinsey report as an autonomous or near-autonomous vehicle. UAVs are supported by Unmanned Aerial Systems (UAS), a complete system of communications and guidance technologies using terrestrial networks and satellites, and their radio frequencies. UAS technology is also, or can be, applied to maritime and terrestrial vehicles. Commercial UAS in the United States, where allowed, currently operates over spectrum assigned to the

Department of Defense and some frequencies assigned for amateur radio operators.[75]

Trains are another example of an autonomous or near-autonomous vehicle. They use guidance systems such as Positive Train Control (PTC), which is designed to use dedicated spectrum to communicate information and guidance to trains.

Advanced Robotics

According to the McKinsey report, robots excel at tasks that require superhuman speed, strength, stamina, or precision in a controlled environment. Advanced robots have greater mobility, dexterity, flexibility, and adaptability. They may have features such as high-definition machine vision and advanced image recognition, and can learn from and interact with humans. Robots can improve efficiency and safety in a variety of tasks as varied as sorting produce, performing high-precision, minimally invasive surgical procedures, or transporting heavy objects. A robot may be stationary, operated on a tether or track, or mobile. Communications requirements vary from environment to environment. A mobile robot operating in open spaces, for example, has different communications and spectrum needs than one controlled within an indoor space. Mobile robots operating with wireless sensors and communications connections have the potential to perform dangerous tasks that potentially jeopardize human life or health, including military combat.

Some advanced robotic systems incorporate existing wireless technologies that have spectrum allocations on unlicensed frequencies used for Wi-Fi or RFID, or over infrared electromagnetic spectrum.

Policy Considerations

Some of the barriers to deploying new technologies are described in detail in the McKinsey report. Briefly, some of the main policy concerns—in addition to spectrum access—deal with

- Privacy.
- Data security.
- Education and training.
- Shifts in labor markets.

- Interaction of state and federal laws and regulations.
- Research and development.

These topics are outside the scope of this report but may need to be addressed in order to provide a hospitable environment for innovation in mobile technology.

End Notes

[1] Broadband delivers large amounts of data at high speeds. The IP-technology currently being deployed for broadband is Long Term Evolution (LTE), which is categorized as a fourth-generation (4G) wireless technology. The next version of LTE, known as LTE Advanced is now being deployed in the United States, requiring new investment in infrastructure. Information about LTE, at http://www.4gamericas.org/index.cfm?fuseaction=page&sectionid=249; about LTE Advanced, at http://www.4gamericas.org/index.cfm?fuseaction=page&sectionid=352.

[2] The White House, Office of Science and Technology Policy & The National Economic Council, *Four Years of Broadband Growth*, June 2013.

[3] Federal Communications Commission, *Connecting America: The National Broadband Plan*, March 17, 2010, http://download.broadband.gov/plan/national-broadband-plan.pdf.

[4] Presidential Memorandum: Unleashing the Wireless Broadband Revolution, June 28, 2010.

[5] Key NTIA actions are discussed in CRS Report R42886, *The National Telecommunications and Information Administration (NTIA): Issues for the 113th Congress*, by Linda K. Moore.

[6] Key provisions in the act are discussed in CRS Report R43256, *Spectrum Policy: Provisions in the 2012 Spectrum Act*, by Linda K. Moore.

[7] For purposes of allocation and assignment, spectrum is segmented into bands of radio frequencies measured in cycles per second, or hertz. Standard abbreviations for measuring frequencies include kHz kilohertz or thousands of hertz; MHz megahertz, or millions of hertz; and GHz—gigahertz, or billions of hertz. The designation can refer to an entire band, such as the 5 GHz band, or to specific frequencies..

[8] U.S. House of Representatives, Committee on Energy and Commerce, Subcommittee on Communications and Technology, Background Memos and Hearings, "Equipping Carriers and Agencies in the Wireless Era," June 27, 2013; "Challenges and Opportunities in the 5GHz Spectrum Band," November 13, 2013; and "Oversight of the Federal Communications Commission, May 20, 2014. Also, U.S. Senate, Committee on Commerce, Science, and Transportation, Subcommittee on Communications, Technology, and the Internet, Hearing, "The State of Wireless Communications," June 4, 2013. The Committee of Energy and Commerce has issued a white paper entitled "Modernizing U.S. Spectrum Policy," which lays out a series of questions about spectrum policy that Congress may act to address, April 1, 2014; link at http://energycommerce.house.gov/press-release/committee-releases-commactupdatewhite-paper-focused-spectrum-policy.

[9] The Omnibus Budget Reconciliation Act of 1993 (P.L. 103-66) amended the Communications Act of 1934 with a number of important provisions affecting the availability of spectrum. The act laid out the general requirements for the FCC to establish a competitive bidding methodology and consider, in the process, objectives such as the development and rapid

deployment of new technologies. (47 U.S.C. §309 (j), especially (1), (3), and (4).) The law prohibited the FCC from making spectrum allocation decisions based "solely or predominately on the expectation of Federal revenues." (47 U.S.C. §309 (j) (7) (A).)

[10] U.S. House of Representatives, Committee on Small Business, hearing, "Building on the Wireless Revolution: Opportunities and Barriers for Small Firms," February 11, 2014.

[11] McKinsey Global Institute, *Disruptive Technologies: Advances that will transform life, business and the global economy*, May 2013, available at http://www.mckinsey.com /insights/business_technology/disruptive_technologies. Note that projections are for global growth and may not accurately represent a technology's role in the U.S. economy.

[12] See, for example, "A Bruising Fight for Survival; mobbed by mobile devices, big players in PCs struggle to adapt," by Nick Wingfield, *The New York Times*, July 29, 2013.

[13] See, for example, "DARPA Seeks Clean-Slate Ideas for Mobile Ad Hoc Networks (MANETS)," DARPA, April 30, 2013 at http://www.darpa.mil/NewsEvents/Releases /2013/04/30.aspx; National Science Foundation Future Internet Project http://www.nets-fia.net/ and Named Data Networking http://named-data.net/.

[14] Testimony of Randal S. Milch, Executive Vice President and General Counsel, Verizon Communications Inc., U.S. Senate, Committee on the Judiciary, Subcommittee on Antitrust, Competition Policy, and Consumer Rights, "An Examination of Competition in the Wireless Market," February 26, 2014.

[15] Thomas Gryta, "U.S. Cellular, NTELOS in Focus After Leap Wireless Buy," *Wall Street Journal*, July 15, 2013.

[16] For example, 3GPP created separate standards for two band classes in the 700 MHz band. After Auction 73, for commercial licenses in the 700 MHz band, small carriers were concentrated in a less-favored band class in the lower half of the 700 MHz band. These carriers complained to the FCC that manufacturers were not providing devices for their band class because of the limited size of the market. They requested the FCC to mandate interoperability across all band plans in the lower 700 MHz band. Among the countervailing arguments against interoperability was that it would undermine future investments in 3GPP standards. An agreement was reached for a voluntary industry solution. In October 2013, the FCC adopted a Report and Order and Order of Proposed Modification to implement the agreement, http://www.fcc.gov/document/700-mhz-interoperability.

[17] For example, public comments regarding service rules for spectrum licenses in the 700 MHz band that were auctioned in 2008 (Auction 73) revealed significant differences between wireless carriers and the wireless technology industry. The latter argued—with little success—for rules that would increase spectrum access for new entrants and new businesses.

[18] Information about 3GPP is at http://www.3gpp.org/about-3gpp. 3GPP is represented in the United States by 4G Americas, http://www.4gamericas.org/index.cfm?fuseaction=home. The mission of 4G Americas is "to promote, facilitate and advocate for the deployment and adoption of the 3GPP family of technologies" and "to develop the expansive wireless ecosystem of networks, devices, and applications enabled by GSM and its evolution to LTE."

[19] Information about ETSI at http://www.etsi.org/about/introduction.

[20] ATIS and the Telecommunications Industry Association (TIA) are global standards development groups based in the United States. Information about ATIS at http://www.atis.org/. Information about TIA at http://www.tiaonline.org/ about/.

[21] About 70% of the approximately 400 members of 3GPP are represented as members of ETSI. ATIS represents about 8% of members, including the top four U.S. wireless carriers. The

balance of members are represented through Asian standards groups. See http://www.3gpp.org/about-3gpp/membership.

[22] Information about the Wi-Fi Alliance at http://www.wi-fi.org/.

[23] As required by the Spectrum Act, the FCC commenced a proceeding to identify new unlicensed spectrum capacity at 5 GHz, http://www.fcc.gov/document/5-ghz-unlicensed-spectrum-unii. See also CRS Report R43256, *Spectrum Policy: Provisions in the 2012 Spectrum Act*, by Linda K. Moore.

[24] Using the International Electrical and Electronic Engineers standard, IEEE 802.11ac, see http://www.ieee.org/ index.html. Also, link to tutorial at http://www.radio-electronics.com/info/wireless/wi-fi/ieee-802-11ac-gigabit.php.

[25] See http://www.itu.int/en/about/Pages/default.aspx. The Government Accountability Office notes that "The federal government considers ITU the principal, competent, and appropriate international organization for the purpose of formulating international treaties and understandings regarding certain telecommunications matters." *Better Coordination and Enhanced Accountability Needed to Improve Spectrum Management*, GAO-02-906, September 2003, p. 19, fn. 26.

[26] The standards developed by the Groupe Special Mobile (GSM) were first adopted by European Union member countries; deployment began in 1998. The GSMA global membership is comprised of GSM and LTE network operators; http://www.gsma.com/aboutus/history.

[27] The next WRC, WRC-2015, is scheduled to occur in November 2015; http://www.itu.int/en/ITU-R/conferences/wrc/ 2015/Pages/default.aspx. Separate tracks of preparations to develop the U.S. positions on WRC agenda items are handled by the FCC and the NTIA. The Office of Spectrum Management of NTIA, in consultation with federal agencies, reviews the WRC agenda and prepares its comments for the U.S. position. NTIA and the FCC solicit input from the private sector and create working groups to address specific agenda items. NTIA and the FCC submit recommendations to the Department of State. The Department of State coordinates and mediates the development of the U. S. position for each WRC and leads the U.S. delegation at each conference.

[28] For some examples, see http://ec.europa.eu/digital-agenda/en/rspp-roadmap-wireless-europeand http://ec.europa.eu/ digital-agenda/.

[29] Small cells are low-powered radio access nodes that are used to boost capacity and manage network interference and connectivity. The types of small cells are Femtocells, typically used in a home; Picocells, that may serve a business; Metrocells, for urban areas; and Microcells, the largest in terms of geographic coverage, used primarily in rural areas.

[30] High site refers to antenna placed on cell towers or other tall structures, to maximize coverage in a large geographic area. Low site indicates that the placement of antennas for small cells is not dependent on height to provide coverage.

[31] A projection of capital expenditures prepared by Mobile Experts shows investment in macro networks plateauing in 2014 and then trending downward while investment in small cell infrastructure continues to rise. The data might be interpreted to indicate that the trend lines will cross as early as 2020. "Mobile Infrastructure Trends," May 2014.

[32] A set of techniques to manage network resources; predictability is an example of a service quality attribute.

[33] Paul Taylor, "AT&T Shifts to 'Virtualised' Hardware," *Financial Times*, February 26, 2014.

[34] "Interference Limits Policy," White Paper, FCC Technical Advisory Committee, Receivers and Spectrum Working Group, February 6, 2013, http://transition.fcc.gov/bureaus/oet/tac /tacdocs/WhitePaperTACInterferenceLimitsv1.0.pdf.

[35] Byoung-Jo J. Kim and Paul S. Henry, "Directions for future cellular mobile network architecture," First Monday: Peer-Reviewed Journal on the Internet, December 3, 2012, http://firstmonday.org/ojs/index.php/fm/article/view/4204.

[36] James Middleton, "Small Cells Should Be Seen as Businesses, Not Engineering Solution," telecoms.com, February 12, 2014; and Dan Jones, "SpiderCloud Eyes LTE Enterprise Small Cells in 2014," News Analysis, LightReading, January 17, 2014.

[37] Examples include DARPA's Mobile Hotspots Program that will retrofit drones to provide Wi-Fi to remote areas; Facebook's plans to develop drone technology to assure global Internet connectivity; and Google's acquisition of drone-maker Titan Aerospace.

[38] A presentation on small cell technology by Alcatel Lucent provides a good picture of scalability and the use of micro networks for LTE, FCC Workshop on 3.5 GHz Spectrum Access System, January 14, 2014, http://wireless.fcc.gov/ workshop/OVERVIEW%20-%20Milind%20Buddhikot%20-%20Alcatel%20Lucent.pdf.

[39] Oceus Networks uses a similar concept to provide communications on the move for the Department of Defense. See http://oceusnetworks.com/.

[40] The Small Cell Forum has published a report on the business case for small cell deployments: http://www.scf.io/en/documents/098_-_Urban_small_cells_in_the_real_world_case_studies.php.

[41] Public discussions of 5G tend to be general in part because of the rapidly evolving state of the technology and its proprietary nature.

[42] "SK Telecom Starts Joint Research with Ericsson for 5G," SK News press release, July 7, 2014, http://www.sk.com/ Channel/News/view/1219.

[43] "DoCoMo to Conduct 5G Experiment Trials with World-Leading Mobile Technology Vendors," NTT DOCOMO press release, May 8, 2014. The Japanese carrier reportedly is planning to introduce 5G service by 2020, offering 1,000 times the capacity of LTE, https://www.nttdocomo.co.jp/english/info/media_center/pr/2014/0508_00.html.

[44] "Ericsson 5G Delivers 5Gbps Speeds," Ericsson News Center, July 1, 2014, http://www.ericsson.com/news/ 1810070.

[45] In China 5G research and development is being sponsored by three ministries that jointly established IMT-2020 (5G) Promotion Group, http://www.itu.int/dms_pub/itu-r/oth/0a/06/R0A0600005D0001PDFE.pdf.

[46] Based at the University of Surrey in the United Kingdom, http://www.surrey.ac.uk/5gic/about/.

[47] Mobile and wireless communications Enablers for the Twenty-twenty Information Society, a European Union project, https://www.metis2020.com/.

[48] Based at the University of Surrey, in the United Kingdom, http://www.surrey.ac.uk /5gic/about/.

[49] The 5G Infrastructure Public-Private Partnership was initiated by the European Union to partner with private sector companies, http://5g-ppp.eu/.

[50] For a more detailed explanation of carrier aggregation, see 3GPP, "Carrier Aggregation Explained" at http://www.3gpp.org/technologies/keywords-acronyms/101-carrier-aggregation-explained.

[51] A discussion of exponential growth as disruptive technologies mature is presented in Erik Byrnjolfsson and Andrew McAfee, *The Second Machine Age: Work, Progress and Prosperity in a Time of Brilliant Technologies*, W.W. Norton and Company Ltd., January 20, 2014.

[52] Qualifying designated entities bidding for licenses in a spectrum auction receive a credit against the purchase price of a successful bid. Current FCC rules provide for three categories of DE, based on gross revenue: small business, very small business and

entrepreneur. Small business and entrepreneur bidders receive a credit of 15%; very small businesses, 25%. (47 C.F.R. 27.1218)

[53] The FCC has been assigning area-based licenses since 1982. A description of the geographic coverage for each type of license is published by the FCC's Office of Engineering and Technology, "FCC Areas," at http://transition.fcc.gov/ oet/info/maps/areas/.

[54] In an auction of EA licenses for frequencies in what is known as the H Block (Auction 96, completed February 27, 2014), Dish Network outbid the competition to acquire all 176 licenses on offer.

[55] "FCC Adopts Rules for First Ever Incentive Auction," FCC, *Report and Order*, Docket No. 12-268, released June 2, 2014 http://www.fcc.gov/document/fcc-adopts-rules-first-ever-incentive-auction-0.

[56] P.L. 112-96, §6402,126 STAT. 224.

[57] "Winning Bidder Announced for Auction 96," FCC Public Notice, DA 14-279, February 28, 2014, http://transition.fcc.gov/Daily_Releases/Daily_Business/2014/db0228/DA-14-279A1.pdf.

[58] FCC, Auction of Advanced Wireless Services Licenses Scheduled for November 13, 2014; Comment Sought on Competitive Bidding Procedures for Auction 98, Docket No. 14-78, May 19, 2014, http://transition.fcc.gov/ Daily_Releases/Daily_Business/2014/db0520/DA-14-669A1.pdf.

[59] Disbursements from the Public Safety Trust Funds are detailed in CRS Report R42543, *The First Responder Network (FirstNet) and Next-Generation Communications for Public Safety: Issues for Congress*, by Linda K. Moore.

[60] FCC Docket GN 12-268; the FCC provides information on the auction preparations at http://wireless.fcc.gov/ incentiveauctions/learn-program/.

[61] "Wireless Telecommunications Bureau Seeks Comment on a Proposal to License The 600 MHz Band Using "Partial Economic Areas,", DA 13-2351, December 11, 2013; and "WTB Provides Details about Partial Economic Areas," Docket No. 12-268, released June 2, 2014 http://www.fcc.gov/document/wtb provides details about partial economicareas.

[62] "FCC Adopts Revised FCC Mobile Spectrum Holdings Policies," FCC News, May 15, 2014, http://www.fcc.gov/ document/fcc-adopts-revised-mobile-spectrum-holdings-policies and *Report and Order,* Docket No. 12-269, released June 2, 20014, http://www.fcc.gov/document/mobile-spectrum-holdings-report-and-order.

[63] The major auction of licenses for the 700 MHz band was Auction 73. Some information about the top ten successful bidders in Auction 73 is available at Wireless Strategy, FCC Auctions, http://www.wirelessstrategy.com/ 700auction.html.

[64] "Fact Sheet FCC Mobile Spectrum Holdings," May 2014, FCC staff report, http://www.fcc.gov/document/fact-sheetfcc-mobile-spectrum-holdings-rules.

[65] U.S. House of Representatives, Subcommittee on Communications and Technology, Committee on Energy and Commerce, Hearing, "Oversight of the Federal Communications Commission," May 20, 2014; written Testimony of Tom Wheeler, Chairman, Federal Communications Commission.

[66] Ibid, p. 2.

[67] FCC, *Further Notice of Proposed Rulemaking,* GN Docket No. 12-354, released April 23, 2014.

[68] WiMax was an early standard for 4G that has largely been displaced by LTE.

[69] A report from the Organization for Economic Co-operation and Development covers many of the issues discussed in this section. It provides information on shared spectrum regulation, standards, and harmonization from a global perspective; OECD (2014), "New Approaches

to Spectrum Management," *OECD Digital Economy Papers*, No. 235, OECD Publishing, http://www.oecd-ilibrary.org/science-and-technology/new-approaches-to-spectrum-management_5jz44fnq066c-en.

[70] 47 U.S.C. §308 (j) (8) requires that net proceeds from competitive bidding for spectrum licenses be deposited in the U.S. Treasury. Net proceeds are the auction revenues minus the FCC's expenses. In addition to the Spectrum Act, Congress has twice in the past amended the provision in order to use auction proceeds for other purposes by creating special funds to hold and disburse auction proceeds. The Commercial Spectrum Enhancement Act, Title II of P.L. 108- 494, created the Spectrum Relocation Fund; the Deficit Reduction Act of 2005 created the Public Safety and Digital Television Transition Fund.

[71] For example, see remarks made by a panel of experts in a forum organized by New America Foundation, "Implementing the PCAST Spectrum Sharing Report; A Citizens Broadband Service and Beyond," April 4, 2014.

[72] McKinsey Global Institute, *Disruptive Technologies: Advances that will transform life, business and the global economy*, May 2013, available at http://www.mckinsey.com /insights/business_technology/disruptive_technologies. Note that projections are for global growth and may not accurately represent a technology's role in the U.S. economy.

[73] See, for example, "Remarks of Ruth Milkman, Chief, Wireless Telecommunications Bureau, FCC," prepared for delivery at Georgetown Center for Business and Public Policy and PCCA Workshop, June 14, 2013.

[74] The Federal Highway Administration and other agencies within DOT are preparing for technology concept pilots. The Connected Vehicle Pilot Deployment Program will test innovations in connected vehicle and mobile device technologies with one or more pilots in 2015, https://www.federalregister.gov/articles/2014/03/12/2014-05414/ connected-vehicle-pilot-deployment-program-request-for-information

[75] As required by Congress, the FAA has taken steps to accommodate the development of commercial UAS operations in the United States. For example, it has created a road map to plan for and document future UAS uses. Six sites have been chosen to test the possibility of allowing commercial drones and aircraft to share airspace, http://www.faa.gov/ about/initiatives/uas/.

In: Spectrum Management and Policy in the U.S. ISBN: 978-1-63482-265-7
Editor: Wanda Flores

Chapter 3

MODERNIZING U.S. SPECTRUM POLICY*

House Energy and Commerce Committee

The Committee on Energy and Commerce is issuing a series of white papers As the first step toward modernizing the laws governing the communications and technology sector. The primary body of law regulating this sector was passed in 1934 and has not been substantially updated in 18 years. Changes in technology and the rate at which they are occurring warrant an examination of whether, and how, communications law can be rationalized to address the 21st century communications landscape. For this reason, the committee initiated an examination of the regulation of the communications industry, and seeks comments from all interested parties on the future of the law.

HISTORY OF SPECTRUM ALLOCATION, REGULATION, AND LICENSING

Spectrum allocation, regulation, and licensing began well before the Communications Act of 1934 ("the Act"). Prior to 1934, the Radio Act of 1912, which vested licensing authority with the Secretary of Commerce, regulated radio. Passed in the wake of the Titanic disaster, the Radio Act of 1912 sought to regulate wireless communications to prevent interference from stifling the potential uses of radio. Just fifteen years later, Congress passed the

* This is an edited, reformatted and augmented version of a white paper issued April 1, 2014.

Radio Act of 1927, which moved licensing authority from the Secretary of Commerce to the newly created Federal Radio Commission. Finally, in 1934 Congress passed the Act, creating the Federal Communications Commission (FCC), which continues to be the primary regulatory body for spectrum regulation today.

Licensing. Since the 1927 Act, the electromagnetic spectrum has been a public good licensed for private use. However, the mechanisms for licensing and rules that surround use of the license have evolved significantly. Under the 1912 Act any American could apply for a license to use the airwaves and the Secretary of Commerce was compelled to grant a license for use. In the 1927 Act, that all changed, with the Federal Radio Commission empowered to choose among competing applications, and select the one that it deemed most in the "public interest." While this was one possible way to solve the problem of scarcity, the practical reality of this system was not lost on the newly formed FRC: some applicants would be told by the government, "there is no room for you."[1]

Starting in 1934, the FCC began to use hearings to determine to whom it would license spectrum. For the first 10 years of the FCC, this was a relatively straightforward process, as there were no instances of multiple applications for the same license. However, in 1944 two applicants sought licenses for the same band of spectrum. These mutually exclusive applications led to the Supreme Court case of Ashbacker Radio Corp. v. FCC,[2] which required the FCC to consider mutually exclusive applications in a single comparative hearing, setting the stage for nearly 40 years of what the industry affectionately called "beauty contests."

Recognizing the shortcomings of the comparative hearing process, in 1982 Congress authorized the FCC to use random selection (lotteries) to resolve mutual exclusivity. While this system had some notable uses in the 1980s,[3] lotteries were all but eliminated by Congress in 1997.[4] While the technology migrated from broadcast radio, to broadcast television, to the earliest mobile phones, the absolute power of the FCC to determine who it thought would put the spectrum to the best use continued unabated until 1993.

In 1993, Congress amended the Communications Act[5] to create Commercial Mobile Radio Services (CMRS)—what many commonly call cellular or wireless services—and authorized the FCC to conduct auctions for certain spectrum licenses. Following on early auction successes, Congress again responded in the Balanced Budget Act of 1997,[6] expanding the FCC's auction authority to cover most types of spectrum licenses.

Finally, in 2012, Title VI of the Middle Class Tax Relief and Job Creation Act (Spectrum Act) expanded the Commission's spectrum auction authority, authorizing the Commission to conduct voluntary incentive auctions: auctions designed to provide an economic incentive for current licensees to relinquish spectrum in exchange for compensation. The law grants the Commission authority for a one-time, specialized incentive auction in which broadcast television stations may relinquish spectrum for Commission auction, as well as ongoing authority to hold other incentive auctions.

Discussion and Questions

1) As discussed in white paper #1 on Modernizing the Communications Act, the telecommunications industry has experienced a great deal of convergence in recent years. One result is that the current licensing structure at the FCC may no longer be the most efficient or appropriate method to maximize spectrum use. The FCC is responsible for licensing spectrum for a number of services, including public safety, fixed and mobile wireless, broadcast television and radio, and satellite. Although many of the processes are the same among these services, the licensing authority is housed in disparate bureaus. What structural changes, if any, should be made to the FCC to promote efficiency and predictability in spectrum licensing?
2) Spectrum users are allowed to operate without an FCC license—subject to certain technical rules—in spectrum that is designated as "unlicensed." In 1985, the FCC opened up frequency bands, including the 2.4 GHz band, for unlicensed communications, and has since allocated other bands specifically for unlicensed operators. Users of unlicensed spectrum do not have exclusive use rights and are subject to interference by others. While operators do not need a license, they must abide by other regulatory safeguards, including authorization of equipment, accepting any interference and not causing harmful interference to others, and ceasing operations upon FCC notification. There is vigorous debate over the appropriate role for unlicensed spectrum in the wireless ecosystem, particularly following the passage of the Spectrum Act. The Act requires the FCC to auction all spectrum made available by the incentive auction, but allows for unlicensed use in guard bands. Some contend that there is an ample amount of unlicensed spectrum available and that assigning spectrum

via exclusive licensing is the most effective, efficient, and economically responsible way to allocate spectrum. Others argue that repurposed spectrum should be allocated for unlicensed use for similar reasons. What role should unlicensed spectrum play in the wireless ecosystem? How should unlicensed spectrum be allocated and managed for long-term sustainability and flexibility?

3) Spectrum sharing is one proposed technological solution that addresses the issue of spectrum scarcity and encourages efficiency. There are multiple ways to share spectrum, including geographic sharing, temporal sharing, and sharing through dynamic spectrum access. In July 2012, the President's Council of Advisors on Science and Technology (PCAST) issued a report on ways to realize the full potential of government held spectrum. The report concluded that sharing is the most efficient way to utilize spectrum and directed the Secretary of Commerce to immediately identify 1,000 MHz of federal spectrum for shared use. However, others assert that spectrum sharing is only part of the solution to spectrum scarcity and that clearing unused or underused federal for exclusive commercial use is a vital part of any strategy for maximizing spectrum resources. In order to enable this sort of reallocation, bipartisan legislation has been introduced in the House that would allow government spectrum users an option to relinquish spectrum and receive a portion of net auction revenues instead of relocation costs, a structure similar to that of the broadcast television spectrum incentive auctions. What should be done to encourage efficient use of spectrum by government users?
4) Given the enormous economic benefits of innovation spurred by commercial spectrum availability, both the government and the private sector are concerned with making more spectrum available to meet commercial demand. When discussing available resources, the FCC considers spectrum to be "currently available" if providers have the legal authority to build out and provide services using that band, or "in the pipeline" if it is not currently available for commercial services but there are government plans to make it available to commercial providers within the next three years. Congress and the FCC have worked to increase the amount of spectrum available to commercial providers, including through the provisions for auctions and relocation in the Middle Class Tax Relief and Job Creation Act. What other steps can be taken to increase the amount of commercially available spectrum?

5) In order to issue spectrum licenses, the Communications Act requires the FCC to make an affirmative finding that granting the license serves the public interest, convenience, and necessity. Moreover, the Act prohibits the FCC from basing its finding on the expectation of auction revenues. Should the Act permit the FCC to use expected auction revenue as the basis for a public interest finding? What criteria should the FCC consider as part of its analysis?
6) The FCC's existing process manages spectrum use through allocation and assignment— bands are allocated for specific services or classes of users, and licenses for use of specific portions of spectrum are assigned to entities. Many of the existing allocations were made because certain spectrum bands are better suited for certain uses. However, changes in technology have changed assumptions over the years. While restrictions have eased in recent years, there are still certain limited-use spectrum licenses. Flexible use licenses permit licensees to use their spectrum for any service, including wireless, broadcast, or satellite services. Should all FCC licenses be flexible use? In what instances should the Commission exercise control over the service offered? How can the Act enable better use of spectrum, either flexible or specified? Finite supply and ever increasing demand have created the scarcity around which the FCC's regulatory controls are based. The FCC has placed limitations on spectrum holdings in a number of ways. In mobile wireless, the Commission has implemented policies that included the cellular cross-interest rule, the Personal Communications Service (PCS) cross-ownership rule, and the Commercial Mobile Radio Services spectrum cap. Currently, the Commission conducts a case-by-case analysis of spectrum aggregation for each entity. The two-part "spectrum screen" first analyzes changes in market concentration that would result from the proposed transaction, and then examines the amount of spectrum that is suitable and available on a market-by-market basis. Prompted by the passage of the Middle Class Tax Relief and Job Creation Act, the FCC initiated a proceeding to review existing policies regarding mobile spectrum holdings to determine whether they still satisfy the statutory goals of promoting competition and avoiding excessive concentration of licenses, given changes in technology, spectrum availability, and the overall marketplace.
7) The FCC has considered other tools to try and enhance competition within the wireless services market. Among these are spectrum "set-

asides," where blocks of spectrum are reserved for a particular type of bidder; bidding credits, which provide a discount on winning bids to small businesses or to specific groups like women and minorities to encourage bidding; and auction design, including reserve prices, package bidding, and proposed restrictions on bidder eligibility. Given the complexity of spectrum auctions, these policies have been criticized for altering the playing field and distorting outcomes. What principles should Congress and the FCC consider when addressing spectrum aggregation limits? How has the converging marketplace and growing demand for services changed the discussion of spectrum aggregation?

8) The FCC further promotes efficient use of spectrum through the build-out requirements and operating rules attached to licenses. Build-out rules require licensees to construct and activate infrastructure within a certain timeframe, or risk losing that license. The operating rules require some licensees to return a license if not used for any 12-month period after construction, promoting the active and continual use of spectrum. These provisions help to ensure that spectrum that is not fully utilized becomes available to those who will put it to dynamic use. Should the Act promote competitive and efficient use of spectrum in this way? How effective is the current Act in doing so? How effectively has the FCC used the tools at its disposal to encourage competition?
9) As discussed above, interference can pose a major problem to efficient and full use of spectrum by providers. The FCC sets limits on transmissions, but doesn't regulate the receivers used by wireless devices to receive wanted signals and eliminate the noise coming from the other surrounding spectrum bands. Underperforming receivers can prevent a device from operating properly. While the FCC has used tools like guard bands to mitigate the potential for interference, recent examples of receiver overload have shown that these efforts may not be enough as demand for spectrum increases but resources become more and more constrained. Some have proposed receiver standards as a solution, but others argue that such a step could result in over-engineering and higher consumer prices. What is the best balance between mitigating interference concerns and avoiding limiting flexibility in the future? Can engineering and forward-looking spectrum strategies account for the possibility of unanticipated technologies and uses in adjacent spectrum bands? How do we

promote flexibility without unreasonably increasing the cost of services and devices? Does the Act provide the FCC tools to address this problem?

10) The other governing body of domestic spectrum use is the National Telecommunications and Information Administration (NTIA), which has the authority to assign spectrum frequencies to all federal government owned or operated radio stations under section 305 of the Communications Act. NTIA manages the federal government's use of spectrum, in coordination with the FCC. Distinctions between "federal" or "non-federal" bands of spectrum are administrative creations made through agreements between the FCC and NTIA. The Spectrum Act required NTIA to work with the FCC to identify specific bands for release to commercial use and how to repurpose resources from federal to commercial use, with priority given to options that assign spectrum for exclusive, non-federal use through competitive bidding. In a report on reducing duplication in the federal government, GAO identified spectrum management as 'fragmented' between NTIA and the FCC and urged coordination.7 What role should NTIA play in the licensing and management of spectrum? Is their current role appropriate and necessary, given the potentially duplicative functions of the FCC and NTIA in spectrum allocation and assignment?

While these questions address specific spectrum issues, the committee will encourages comment on any aspect of spectrum policy. Please respond by April 25, 2014, to commactupdate@mail.house.gov. For additional information, please contact David Redl at (202) 225-2927.

End Notes

[1] Federal Radio Commission, First Annual Report, at 6 (1927).

[2] 326 U.S. 327 (1945).

[3] See, e.g. In re: Selection by Lottery for Competing Cellular Applications, 98 F.C.C. 2d 175 (1984) (use of lotteries to distribute licenses for the cellular service in markets other than the thirty largest).

[4] Pub. L. No. 105-33, Title III, 111 Stat. 251 (1997) (Balanced Budget Act of 1997).

[5] Pub. L. No. 103-66, Title VI, 6002(b), 107 Stat. 312 (1993) (Omnibus Budget Reconciliation Act of 1993 ("OBRA-1993")).

[6] Balanced Budget Act of 1997.

[7] U.S. Government Accountability Office, GAO-12-342SP: 2012 Annual Report: Opportunities to Reduce Duplications, Overlap and Fragmentation, Achieve Savings, and Enhance Revenue 89 (2012).

In: Spectrum Management and Policy in the U.S. ISBN: 978-1-63482-265-7
Editor: Wanda Flores

Chapter 4

SPECTRUM MANAGEMENT: FCC'S USE AND ENFORCEMENT OF BUILDOUT REQUIREMENTS*

United States Government Accountability Office

WHY GAO DID THIS STUDY

Radio frequency spectrum is a natural resource used to provide a variety of communication services, such as mobile voice and data. The popularity of smart phones, tablets, and other wireless devices among consumers, businesses, and government users has increased the demand for spectrum. FCC takes a number of steps to promote efficient and effective use of spectrum. One such step is to establish buildout requirements, which specify that an entity granted a license must begin using the assigned spectrum within a specified amount of time or face penalties, such as loss of the license.

GAO was asked to review buildout requirements and the efficient use of spectrum. This report (1) describes the buildout requirements FCC established for wireless services, (2) assesses the extent to which FCC follows its process to enforce buildout requirements, and (3) examines stakeholder opinions on the extent that commonly cited goals for buildout requirements have been met. GAO reviewed FCC regulations and guidance on buildout requirements and

---

* This is an edited, reformatted and augmented version of a United States Government Accountability Office publication, No. GAO-14-236, dated February 2014.

examined FCC license data on outcomes of buildout requirements for 5 out of about 45 wireless services selected to ensure variety in type of use and buildout requirement, among other criteria. GAO also interviewed FCC officials, commercial spectrum licensees, industry associations, and spectrum policy experts.

GAO is making no recommendations in this report. FCC reviewed a draft of this report and provided technical comments that GAO incorporated as appropriate.

What GAO Found

The Federal Communications Commission (FCC) has established buildout requirements—which require a licensee to build the necessary infrastructure and put the assigned spectrum to use within a set amount of time—for most wireless services, including cellular and personal communication services. FCC tailors the buildout requirements it sets for a wireless service based on the physical characteristics of the relevant spectrum and comments of stakeholders, among other factors. Therefore, buildout requirements vary across wireless services. For example, a buildout requirement can set the percentage of a license's population or geographic area that must be covered by service or can describe the required level of service in narrative terms rather than numeric benchmarks. Buildout requirements also vary by how much time a licensee has to meet a requirement and whether it has to meet one requirement or multiple requirements in stages.

FCC's enforcement process for wireless-service licenses with buildout requirements primarily relies on information provided by licensees, and FCC followed its process for the five wireless services GAO reviewed. Specifically, FCC requires licensees to self-certify that they have met buildout requirements. If a licensee does not do so, FCC automatically terminates the license. Some stakeholders GAO interviewed said that self-certification is an effective way for FCC to enforce buildout requirements because it is public and transparent. GAO examined FCC license data for five wireless services and found that buildout requirements were met for 75 percent of those licenses, and FCC generally terminated those that did not. As part of enforcement, FCC also grants or dismisses licensees' requests to extend the deadline for meeting a requirement. FCC may grant an extension if the licensee shows that it cannot meet a deadline due to causes beyond its control, like a lack of available equipment. For the five wireless services examined,

GAO found that extensions were requested for 9 percent of licenses, and FCC granted 74 percent of these requests. FCC officials said that the Commission seeks to be aggressive but pragmatic when enforcing buildout requirements, including being flexible on deadlines when needed. Some licensees and industry associations GAO interviewed said that extensions can provide needed flexibility when unexpected problems occur. Some concerns were raised, however, that granting extensions can undermine buildout requirements by creating an impression that they will not be strictly enforced.

Stakeholders GAO interviewed generally said that buildout requirements are effective in meeting two of four goals commonly cited in FCC documents and statute—encouraging licensees to provide services in a timely manner and preventing the warehousing of spectrum. Stakeholders had mixed views on the effectiveness of buildout requirements in meeting two other goals—promoting innovative services and promoting services to rural areas—largely because they believed that other tools could better address these goals. Other tools stakeholders mentioned include greater use of spectrum licenses that allow a wider array of uses and providing licensees with subsidies to serve rural areas. Nearly all the licensees and industry associations GAO interviewed said they support FCC having buildout requirements, while spectrum policy experts GAO interviewed were mixed in their support of the requirements. Experts who did not support buildout requirements said that the requirements are set too weak or that other tools could better meet FCC goals, among other reasons.

ABBREVIATIONS

APA	Administrative Procedure Act
EA	economic area
FCC	Federal Communications Commission
GHz	gigahertz
MHz	megahertz
NTIA	National Telecommunications and Information Administration
PCS	personal communications service
ULS	Universal Licensing System
WTB	Wireless Telecommunications Bureau

February 26, 2014

The Honorable Michael F. Doyle
House of Representatives

The Honorable Peter Welch
House of Representative

Radio frequency spectrum is a natural resource used to provide a variety of communication services, such as mobile voice and data, to consumers, businesses, and federal, state, and local government users. The popularity of smart phones, tablets, and other wireless devices has created an explosion in the demand for and use of spectrum. However, most of the usable spectrum in the United States has already been allocated to existing services and assigned to users. Since there are virtually no "green fields" of favorable spectrum currently available to help address the increased demand by the general public and government users, Congress, regulators, and entities that rely on spectrum are looking for ways to promote more efficient use of this resource.[1]

The Federal Communications Commission (FCC)—the agency that regulates spectrum use for commercial and other nonfederal users— manages spectrum through *allocation* and *assignment:*[2]

- *Allocation* involves designating bands of spectrum for specific types of services or classes of users, such as for commercial or government use.[3]
- *Assignment* provides a license to a specific entity, like a wireless carrier or trucking company, to use a specific portion of spectrum after it has been allocated. FCC uses a competitive-bidding process, or auctions, to assign some licenses to entities that submit the highest bids for licensing in specific bands of spectrum.

FCC takes a number of steps in both the allocation and assignment processes to promote efficient and effective use of spectrum. Buildout requirements—also known as construction or coverage requirements— are among the tools that FCC uses to work toward its goal of efficient spectrum use; in its fiscal year 2014 budget request, FCC said that enforcing the Commission's regulations, specifically citing buildout requirements, is one action it takes to achieve its strategic goal to maximize the benefits of spectrum. Buildout requirements specify that an entity granted a license must build the necessary infrastructure and put the assigned spectrum to use within a specified amount of time or face penalties, typically termination of all or part of the license.[4] Buildout requirements are designed to ensure that licensees put spectrum to use within a specific period rather than let it sit idle.

Given the high demand for spectrum, you asked us to review buildout requirements and the efficient use of spectrum. While FCC sets buildout requirements for many types of spectrum licenses, we focused our review on the buildout requirements for wireless services, as these services account for a majority of spectrum licenses—nearly 2 million licenses as of December 2013.[5] This report (1) describes the buildout requirements established by FCC for spectrum licenses for wireless services, (2) assesses the extent to which FCC follows its process to enforce buildout requirements for wireless services, and (3) examines stakeholder opinions on the extent that goals for buildout requirements have been met. To answer these questions, we reviewed FCC regulations and guidance on buildout requirements for services that use spectrum. We also interviewed FCC officials to understand which wireless services have buildout requirements, the general process used to establish buildout requirements, and the steps FCC takes to monitor and enforce buildout requirements for wireless services.

To determine the extent to which FCC enforces buildout requirements for wireless services, we examined FCC data and documents on buildout requirements for 5 out of about 45 wireless services (see table 1). We selected these wireless services to ensure variety in type of service or use, type of buildout requirement, how licenses are assigned (e.g., auctions), and the number of licenses in the service. We also considered recommendations from FCC officials and other interviewees when selecting wireless services.

For the five selected wireless services, we analyzed data from FCC's Universal Licensing System (ULS) to examine license outcomes and FCC enforcement actions related to buildout requirements.[6] For each selected wireless service, we analyzed data for licenses with buildout requirements on or before December 31, 2012.[7] For each license, we examined outcomes related to buildout requirements (e.g., met requirement or did not meet requirement and was terminated) and outcomes related to requesting an extension of a buildout deadline. To assess the reliability of the ULS data, we interviewed FCC officials, reviewed system documentation, and conducted electronic data testing. We determined that these data were sufficiently reliable for our purposes.

Table 1. Description of Selected Wireless Services

Service	Use	Frequency	Assignment process
Broadband personal communications service (PCS)	Mobile voice and data services, including cell phone, text messaging, and Internet	1,850 to 1,990 MHz	Auction
220 MHz phase II	Various uses, including paging services and two-way radio systems	220 to 222 MHz	Auction
39 GHz	Fixed point-to-point communications used, for example, to provide "backhaul"[a] and backbone links for other communications systems; also used for point-to-multipoint systems as part of private wireless networks	38.6 to 40.0 GHz	Auction
Fixed microwave	Fixed point-to-point communications to carry voice, data, video, and other information, often as backhaul[a]	Various frequencies above 928 MHz	Administrative process—assigned on a first-come, first-served basis
Industrial/business private land mobile radio, below 700 MHz (private land mobile radio)	Radio systems used by companies to support business operations like dispatching and remote monitoring of equipment	Various frequencies below 700 MHz, not including 220 MHz	Administrative process—generally assigned on a non-exclusive, shared basis[b]

Source: GAO analysis of FCC information.

[a]"Backhaul" is the telecommunications industry term that refers to connections between a central telecommunications network and other nodes outside the central network, such a cell phone tower.

[b]The private land mobile radio service includes public safety licenses, but we excluded these licenses from our analysis. We focused our review on wireless services licensed by the Wireless Telecommunications Bureau (WTB), which account for a majority of spectrum licenses.

To gather stakeholder opinions on the effectiveness of buildout requirements, we interviewed a sample of 28 industry stakeholders including spectrum policy experts, industry associations, and licensees. We selected experts based on participation in recent GAO spectrum policy reviews, publications on spectrum policy, and recommendations from other interviewees, and we selected industry associations and licensees to cover the five selected wireless services. To determine the goals for buildout requirements, we reviewed applicable law, FCC documents from recent rulemakings, and other FCC budget and performance documents; we identified four goals for buildout requirements that were frequently cited in these source documents. We conducted semi-structured interviews with the 28 stakeholders to gather their opinions on whether buildout requirements were effective or ineffective in meeting the frequently cited goals as well as for reasons or examples to support these opinions. We also obtained stakeholders' views about what changes to buildout requirements and what alternative tools FCC could use to better meet the goals, if any. Findings from our interviews with selected industry stakeholders should not be used to support generalizations about the overall effectiveness of buildout requirements or what alternatives FCC might use. To supplement these interviews, we reviewed filings made by licensees, industry associations, and other companies and associations in two FCC proceedings since 2010 that sought comments on buildout requirements for wireless services. See appendix I for a more detailed explanation of our scope and methodology.

We conducted this performance audit from February 2013 through February 2014 in accordance with generally accepted government auditing standards. Those standards require that we plan and perform the audit to obtain sufficient, appropriate evidence to provide a reasonable basis for our findings and conclusions based on our audit objectives. We believe that the evidence obtained provides a reasonable basis for our findings and conclusions based on our audit objectives.

BACKGROUND

Radio frequency spectrum is used to provide an array of commercial and governmental services, like mobile voice and data, air-traffic control, broadcast television and radio, and public safety activities. In the United States, responsibility for spectrum management is divided between two agencies: FCC and the Department of Commerce's National

Telecommunications and Information Administration (NTIA). FCC manages spectrum use for nonfederal users, including commercial, private, and state and local government users under authority provided in the Communications Act.[8] NTIA manages spectrum for federal government users and acts for the President with respect to spectrum management issues.[9]

FCC is an independent regulatory agency composed of five commissioners appointed by the President and confirmed by the Senate. The commissioners delegate many of FCC's day-to-day responsibilities, including processing applications for licenses and analyzing consumer complaints, to the agency's 7 bureaus and 10 offices. According to its fiscal year 2014 budget request, FCC has just over 1,700 full-time equivalent staff in Washington, D.C., and other locations; FCC requested $359 million for its fiscal year 2014 budget.

Among other duties, FCC bureaus responsible for granting spectrum licenses administer service rules that outline technical and operating requirements for spectrum licenses. Service rules may be set at the time FCC allocates spectrum into bands for a specific type of service or group of users. FCC develops rules through a process defined by the Administrative Procedure Act (APA).[10] The APA process requires FCC to provide the public with notice of its proposed and final rules and with an opportunity to comment as the rules are developed. All comments and information gathered by FCC constitute the public record to support rulemakings and are electronically maintained in a docket. FCC maintains the dockets in an electronic system that is available to the public on its website.[11]

Table 2. FCC Bureaus That License Spectrum

Bureau	Licensing responsibilities
International Bureau	International telecommunications and satellite services.
Media Bureau	Electronic media, broadcast television, and radio.
Public Safety and Homeland Security Bureau	Services used by public safety entities such as police and fire departments.
Wireless Telecommunications Bureau	All domestic wireless services (other than public safety) including cellular telephone, paging, fixed microwave, and mobile broadband services.

Source: FCC.

After spectrum is allocated and service rules are set, depending on the type of service or user, one of four FCC bureaus assigns licenses to users (see table 2).

For example, the Wireless Telecommunications Bureau (WTB) develops and executes policies and procedures for the licensing of all wireless services (except wireless public safety services), and the Media Bureau administers television and radio broadcast licenses.

Licenses for wireless services are assigned through competitive bidding (auctions) or administrative processes.

The assignment process used depends in large part upon whether applications for licenses are mutually exclusive—that is, on whether granting a license to one entity would preclude granting a license to another entity for the same portion of the spectrum in the same geographic location.

- For licenses that are mutually exclusive, FCC typically uses auctions to assign licenses for commercial wireless services. Auctions are a market-based mechanism used to assign a license to the entity that submits the highest bid.[12] In the report, we refer to these licenses as market-based licenses.
- For licenses that are not mutually exclusive, primarily public-safety and private-wireless licenses, FCC generally assigns licenses through administrative processes. For example, FCC distributes some licenses on a first-come, first served basis, where licenses are assigned based on when the license applications were submitted.[13] To maximize the number of spectrum users, FCC often requires license applicants to coordinate. License applicants retain a private third-party firm, known as a frequency coordinator, to select a frequency that minimizes interference to existing licensees. We refer to these licenses as site-based licenses.

FCC uses its ULS database to assign and track licenses for wireless services. ULS operates as a single licensing system used by FCC and licensees to apply for, modify, cancel, and take other actions on licenses for all wireless services in a uniform manner.

FCC Establishes Buildout Requirements for Most Wireless Services

FCC has established buildout requirements for most wireless services, including paging, cellular, land mobile radio, and wireless communications services (see appendix II).[14] FCC officials said that the Commission makes every effort to ensure efficient use of each spectrum license, and, in line with these efforts, FCC uses buildout requirements to help ensure that spectrum is put to use. FCC establishes buildout requirements for wireless services through its rulemaking process, and the buildout requirement for each wireless service is tailored to the particular service. According to FCC officials, they take into account, when setting buildout requirements, (1) stakeholders' comments about the proposed requirements in the notice of proposed rulemaking; (2) the characteristics of the relevant spectrum in terms of propagation of the signal through space and its interaction with obstacles, which could affect the infrastructure costs for the intended coverage;[15] and (3) the types of service in adjacent spectrum, including considerations of harmful interference with those services.

Buildout requirements have three common features, which vary based on the differences in wireless services.

- *Type of requirement.* This refers to the benchmark or outcome that a licensee must meet. There are three types of requirements. A population or geographic coverage requirement sets the percentage of the license's population or geographic area, respectively, that must be covered by service. A construction requirement requires that the system operate consistent with the rules governing the service, specified in the license, by a specific time. Lastly, a "substantial service" requirement describes the level of service that must be provided in narrative terms rather than in absolute, numeric benchmarks, such as with a coverage requirement.[16] When FCC establishes a substantial service requirement, it sometimes includes "safe harbors" in the rulemaking documents. Safe harbors illustrate specific ways that a licensee could demonstrate substantial service for a particular wireless service, such as constructing a certain number of point-to-point links or serving populations that are outside areas served by other licensees.

- *Number of benchmarks.* This refers to whether the licensee must complete the buildout requirement by one deadline, or whether the licensee must complete multiple requirements in stages with corresponding deadlines. When FCC sets more than one benchmark and thus deadline for a license, it refers to the requirements as the interim and final requirements or as the first requirement, second requirement, and so forth.
- *Length of buildout period.* This refers to the length of time from the grant of a license to the buildout deadline or deadlines.

We examined five wireless services, which have buildout requirements that vary in each of the features discussed above (see table 3). Licenses in these five services are subject to all three types of requirements described above. For example, FCC set a substantial service requirement for 39 GHz licenses. In the rulemaking proceeding for 39 GHz, FCC stated that setting a substantial service requirement would permit flexibility in licensees' system design, as the types of possible service vary tremendously and may develop in unpredictable ways. For some services, FCC gave licensees the option to choose the type of requirement. For Broadband PCS, FCC set a population coverage requirement, but also provided the alternative of meeting a substantial service requirement. In terms of the number of benchmarks, FCC set two benchmarks for the Broadband PCS (depending on the license) and 220 MHz services and set a single benchmark for the other three services we selected. The lengths of the buildout period for the five services range from 12 months (for industrial/business private land mobile radio) to 10 years (for all three market-based services).

Under some circumstances, a particular license may not be subject to a buildout requirement, even if FCC established buildout requirements in the wireless-service rules for the license. For the three market-based services we examined, for example, a licensee can divide its license into smaller pieces by disaggregating or partitioning, which divides the assigned spectrum into smaller amounts of bandwidth or smaller geographical areas, respectively. In such cases, some of the resultant licenses may not have buildout requirements because the requirements are met by one of the other pieces of the original license. For example, for Broadband PCS licenses, parties seeking to disaggregate a license must decide which party will be responsible for meeting the buildout requirements or agree to share responsibility for meeting the requirements. Additionally, fixed-microwave and private land-mobile-radio licenses that authorize certain temporary or itinerant use of the spectrum, such as construction work or event planning, would not normally include a buildout requirement, since the license does not permit any long-term or ongoing operations.

Table 3. FCC Buildout Requirements for Selected Wireless Services, as of January 2014

Service	Buildout requirement(s)	Features of buildout requirement(s)
Broadband PCS	30 MHz Licenses: Provide service to at least 1/3 of the population in licensed area within 5 years of initial license grant, and 2/3 of the population within 10 years of initial license grant. Alternatively, licensees may provide substantial service to their licensed area within the appropriate 5- and 10-year benchmarks.	Type: Population coverage requirement, with substantial service alternative Number of benchmarks: 2 Length: 5 and 10 years
	10/15 MHz Licenses: Provide service to at least 1/4 of the population in licensed area within 5 years of initial license grant, or alternatively substantial service within 5 years of initial license grant.	Type: Population coverage requirement, with substantial service alternative Number of benchmarks: 1 Length: 5 years
220 MHz phase II auctioned licenses	Nationwide Licenses: Provide coverage to a composite area of 750,000 square kilometers or 37.5 percent of the U.S. population within 5 years of initial license grant, and 1,500,000 square kilometers or 75 percent of the population within 10 years of initial license grant. Alternatively, nationwide licensees may meet the construction requirements by demonstrating an appropriate level of substantial service at the 5- and 10-year benchmarks.	Type: Population or geographic coveragerequirement, with substantial service alternative Number of benchmarks: 2 Length: 5 and 10 years

Service	Buildout requirement(s)	Features of buildout requirement(s)
	Economic Area (EA) or Regional Licenses: Provide coverage to at least 1/3 of the population in the licensed area within 5 years of initial license grant, and 2/3 of the population within 10 years of initial license grant. Alternatively, EA or Regional licensees may meet their construction requirements by providing an appropriate level of substantial service at 5- and 10- year benchmarks.	Type: Population coverage requirement, with substantial service alternative Number of benchmarks: 2 Length: 5 and 10 years
39 GHz geographic licenses	Provide substantial service to its licensed area within 10 years of initial license grant.	Type: Substantial service requirement Number of benchmarks: 1 Length: 10 years
Industrial/business private land mobile radio below 700 MHz	Must be placed into operation within 12 months of license grant.	Type: Construction requirement Number of benchmarks: 1 Length: 12 months
Fixed microwave	Must be placed in operation within 18 months of initial license grant.	Type: Construction requirement Number of benchmarks: 1 Length: 18 months

Source: GAO analysis and presentation of FCC information.

In other circumstances, licenses for industrial/business private land mobile radio and fixed microwave site-based services can authorize the use of multiple frequencies. For industrial/business private land-mobileradio licenses, for example, the purpose of authorizing multiple frequencies is to improve the efficiency of a multi-user system in which users can use any available channel; it is similar to a multi-lane highway in which cars can use any lane. In such cases, FCC sets a buildout requirement for each frequency authorized by the license. If a licensee fails to construct for a given frequency, FCC automatically terminates the authorization to use that frequency as of the buildout deadline. If all the frequencies for a license are terminated, FCC will terminate the license.[17]

FCC Enforces Buildout Requirements through Automated Processes and Staff Reviews and Followed Its Process for Selected Wireless Services

FCC Enforces Buildout Requirements Primarily in Response to Licensee Actions

To enforce buildout requirements for wireless services, FCC requires licensees to self certify that they met buildout requirements and automatically terminates licenses that fail to do so, in line with FCC rules. Through ULS—the computerized system FCC and licensees use to process and track licenses for all wireless services—licensees submit notifications to inform FCC that a requirement is met.[18] However, if the licensee does not notify FCC that it met a buildout requirement in a timely manner, FCC takes steps through ULS to terminate the license. Specifically, ULS is programmed to automatically carry out steps to terminate a license. Thirty days after a buildout deadline, ULS puts a license into "termination pending" mode. FCC releases a weekly public notice of all the market-based and site-based licenses that entered termination-pending mode. If a licensee does not file a petition for reconsideration within 30 days of the public notice demonstrating that it timely met the requirement, ULS will automatically terminate the license effective as of the buildout deadline.[19] Once terminated, the license is then made available for re-assignment or re-auction.[20]

Beyond the automated steps involving the termination of licenses discussed above, FCC primarily enforces buildout requirements for wireless services by responding to information provided by licensees. In particular, licensees submit information to FCC in ULS through filings, which FCC responds to through automatic processes or staff reviews, depending on the type of filing.[21] While FCC responds to licensee filings, it does not actively monitor licensee performance on buildout requirements for wireless services; that is, FCC does not send teams out to determine the extent of a licensee's buildout. Through ULS, FCC officials told us they have the ability to examine outcomes related to the buildout requirements. While FCC enforces buildout requirements for individual licenses, it does not maintain a comprehensive program that monitors overall licensee compliance with buildout requirements within a service or across services.

Automatic Processes

For licensee filings that do not require evaluation, FCC automates the responses to be carried out in ULS. In general, two specific filings—the required notification for site-based services and request to cancel a license—are automatically reviewed in ULS. According to FCC officials, no formal review is needed for required notifications for site-based licenses because the licensee is certifying that it met the conditions laid out in the license. In addition, a licensee may apply to cancel a license at any time, including before or after a buildout requirement is due. Unless a licensee has other pending applications, an application to cancel a license is automatically approved.

Staff reviews. In contrast, some licensee filings require evaluation, so FCC staff must review these notifications and requests. In particular, FCC staff review is required for notifications for market-based licenses (that is, licenses assigned through auctions), as well as all requests for extensions and requests to accept late-filed required notifications.

- *Required notification.* As with site-based services, a market-based licensee must file a required notification to notify FCC that it met its buildout requirement. Compared to the specific parameters set in site-based licenses, FCC officials said that market-based licenses tend to give licensees more flexibility in how to use spectrum or deploy service; therefore, FCC requires additional documentation—like information on the technology used in a system—to help assess whether a licensee met its buildout requirement. FCC specifies what additional documentation is required in the rules for a wireless service or a public notice. For example, Broadband PCS licensees must submit maps and other supporting documents showing compliance with the 5- and 10- year benchmarks. FCC can also ask a licensee to send additional information if needed to determine if the licensee met the buildout requirement.
- *Request for an extension.* A licensee can also request an extension of the buildout deadline. A request for an extension must be filed before the licensee reaches the buildout deadline for a license. The criteria for when extension requests may be granted are laid out in regulation.[22] For example, FCC will not grant an extension request where delay has been caused solely by a failure to obtain financing. In general, the regulation states that FCC may grant an extension request if the licensee shows that its failure to meet a buildout deadline is due to causes beyond its control. For example, extension requests can be

granted for issues such as a lack of available equipment for a band or interference problems with other spectrum users. FCC staff review each request to determine whether an extension is justified. If an extension request is granted, FCC changes the buildout deadline for a license in ULS, but if the request is dismissed, the original buildout deadline stands. If a licensee still needs additional time after being granted an extension, the licensee can request an additional extension. Beyond individual licensee requests, FCC can grant blanket extensions when warranted for a wireless service or group of licenses. According to FCC officials, FCC has considered a blanket extension for most or all of the licenses in a service in cases where it has observed a relatively high number of extension requests. A licensee or stakeholder may also ask FCC to consider a blanket extension. FCC granted a blanket extension of the 5-year buildout requirement for 220 MHz phase II licenses.[23] When the buildout requirements for these licenses started to come due in 2004, numerous licensees filed extension requests. Licensees and others said that there was insufficient equipment to provide voice communications in the 220 MHz band, so licensees collectively would not be able to meet their buildout requirements. In the order granting the extension, FCC stated that a 3-year extension was warranted because, among other reasons, it would provide time for the equipment market to develop.

- *Waiver or petition for reconsideration.* Licensees can also file requests to submit late-filed required notifications in limited situations. Specifically, a licensee can file a waiver request or petition for reconsideration if it met a buildout requirement but did not file a required notification on time. For example, if the license has entered termination-pending mode, a licensee can file a petition for reconsideration within 30 days of being listed in the weekly public notice to prevent the license from terminating. In the petition, a licensee must provide the date on which it met the buildout requirement, any supporting documentation required by the rules, and the reason the notification was not filed on time. To be granted a waiver, a licensee must submit any required documentation to demonstrate how it met the buildout requirement and must meet the wavier standard set forth in FCC rules.

Figure 1 provides further information on the timing of filings related to buildout requirements.

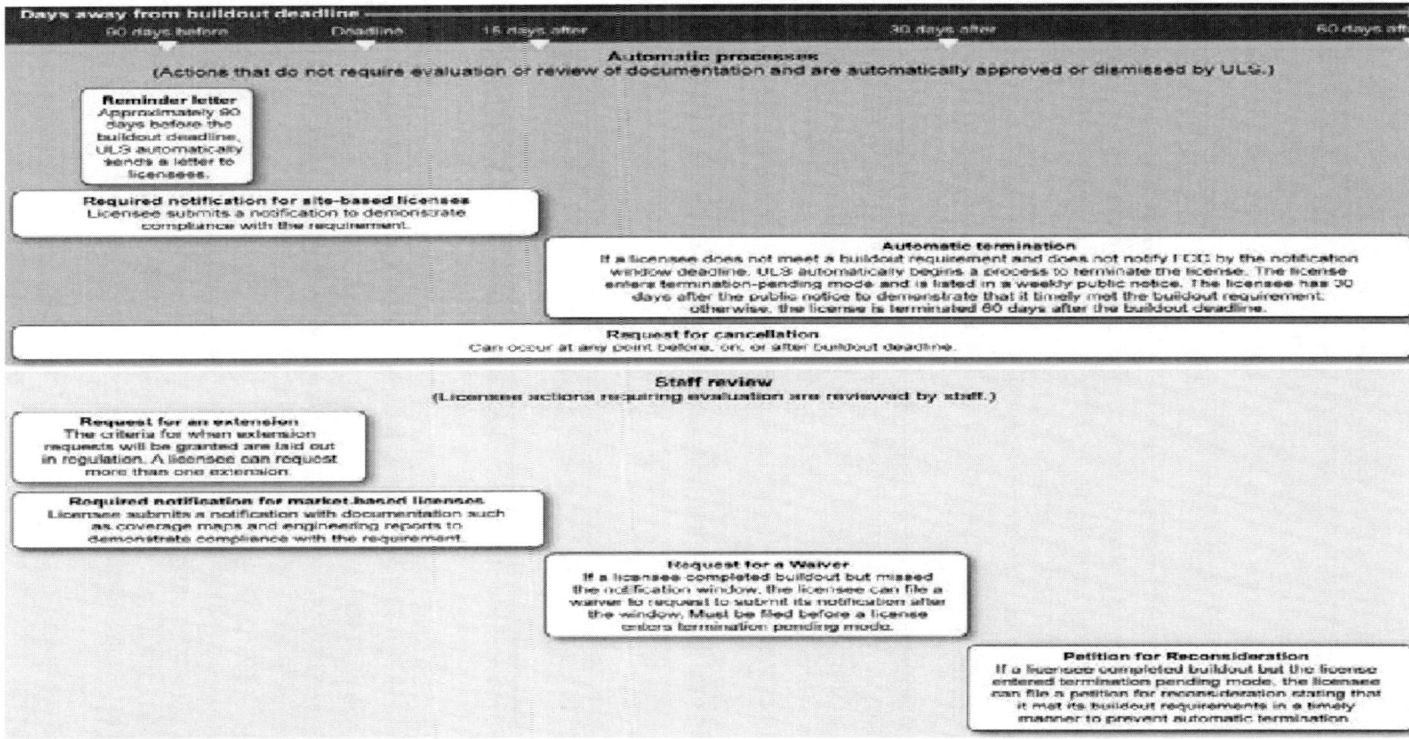

Source: GAO and FCC.

Figure 1. Overview of FCC Enforcement of Buildout Requirements for Wireless Services.

Industry associations and licensees we interviewed generally thought FCC's enforcement process works well. Many stakeholders we interviewed said that FCC's self-certification process is appropriate. For example, one expert, three licensees, and officials from an industry association indicated that the public, transparent nature of the required notifications makes self-certification an effective way to enforce buildout requirements. One expert said that self-certification is the most efficient method for FCC to collect and manage buildout information, as licensees are in the best position to gather and report this information. Moreover, a few licensees and industry associations indicated no other approach would be feasible given the high volume of wireless-service licenses and FCC resource constraints. Furthermore, most of the industry associations and licensees we interviewed said that FCC's ULS system is easy to use. For example, one licensee said that the mechanics of uploading information in ULS for required notifications and requests for extension are straightforward.

FCC Followed Its Enforcement Process for Selected Wireless Services, Although a Few Delays Could Pose Problems for Some Licensees

For the five wireless services we reviewed, buildout requirements were met for many licenses, and when buildout requirements were not met, FCC generally terminated the licenses. Across the five services, we found that buildout requirements were met for 75 percent of licenses (19,582 of 26,217). For 3 of the 5 services we examined, buildout requirements were met for a majority of licenses (see table 4). For the other services, buildout requirements were met for half of the 39 GHz licenses and 19 percent of 220 MHz licenses. When licensees did not meet the buildout requirements, FCC generally proceeded as expected by terminating the licenses. A few of our selected services have a relatively high percent of licenses terminated because licensees did not meet their buildout requirements due to special circumstances. For example, FCC terminated 21 percent of the fixed microwave licenses we examined, but we found that a single licensee held nearly all—1,955 of 2,179—of the terminated licenses.

Table 4. Buildout Requirement Outcomes for Licenses, by Wireless Service, Data as of September 1, 2013

		Buildout requirement not met			
Service	Number (%) of licenses where the buildout requirement was met	Number (%) of licenses terminated[a]	Number (%) of licenses with other outcomes[b]	Number (%) of licenses that did not reach the buildout deadline[c]	Total
Broadband PCS	3,121 (88%)	57 (2%)	22 (1%)	358 (10%)	3,558
220 MHz phase II	181 (19%)	529 (55%)	134 (14%)	119 (12%)	963
39 GHz	839 (50%)	489 (29%)	13 (1%)	331 (20%)	1,672
Fixed microwave	6,299 (60%)	2,179 (21%)	1,388 (13%)	612 (6%)	10,478
Private land mobile radio	9,142 (96%)	326 (3%)	39 (<1%)	39 (<1%)	9,546
All market-based services	4,141 (67%)	1,075 (17%)	169 (3%)	808 (13%)	6,193
All site-based services	15,441 (77%)	2,505 (13%)	1,427 (7%)	651 (3%)	20,024
Total	19,582 (75%)	3,580 (14%)	1,596 (6%)	1,459 (6%)	26,217

Source: GAO analysis of ULS data.

Notes: See appendix I for additional information on the scope and methodology for the analysis. Percentages may total to more than 100 due to rounding.

[a] For the two site-based services (fixed microwave and private land mobile radio), the licenses in the license-terminated column are those for which all frequencies authorized by the license were terminated and thus the license was terminated.

[b] Licenses in the other-outcomes category include those that did not meet their buildout requirement and remained active for some period of time after the buildout deadline.

[c] Licenses in the did-not-reach-buildout-deadline category include those that were (1) canceled on or before the buildout deadline, (2) terminated before the buildout deadline, or (3) expired on or before the buildout deadline.

However, buildout requirements were not met for some licenses, and the licenses were not terminated—that is, licenses in the "other outcomes" category in table 4. In two services—fixed microwave and private land mobile radio—there were mixed outcomes because licenses can authorize use of multiple frequencies. Such licenses can be terminated, in part, with the remainder being active or having no requirements. For example, a licensee could meet the requirement for some frequencies but not others, resulting in the termination of some frequencies but not the license; there were 1,216 fixed-microwave (12 percent) and 13 private land-mobile-radio (less than 1 percent) licenses with this outcome. Also, a licensee could have some frequencies terminated and not have a buildout requirement for other frequencies; 66 fixed-microwave licenses (1 percent) and 17 private land-mobile-radio licenses (less than 1 percent) fit this description.

The remaining licenses in the other-outcomes category are instances where a buildout requirement was not met but a license remained active after the buildout deadline, mostly for good reasons. Examining ULS license data and other FCC documents, we found that there were reasonable explanations—like a licensee canceling a license during the automatic termination process or filing a required notification that FCC has yet to approve or dismiss—for why most of these licenses were not terminated on the buildout deadline. For example, we found that 106 fixed-microwave licenses in the other-outcomes category were canceled during the automatic termination process—that is, within 60 days of the buildout deadline.[24] For the 220 MHz phase II licenses with other outcomes, 110 licenses were canceled during the automatic termination process, and 24 licenses had pending required notifications in ULS. While there were good reasons why most licenses remained active, there were a few instances where ULS contained no explanation. For the Broadband PCS licenses, for example, ULS did not contain information explaining why 16 licenses were not terminated on the buildout deadline.[25] The automatic termination process in ULS did not occur until 2006, after the buildout deadlines for the 16 Broadband PCS licenses. Therefore, FCC officials said that while these licenses were terminated automatically by rule on the deadline, ULS was not updated to reflect the termination for several months after their buildout deadlines.

Lastly, some licenses with buildout requirements did not reach their buildout deadlines; as a result, FCC did not have to enforce the buildout requirements for these licenses. A license could not reach its buildout deadline for three reasons. First, the license could be canceled by the licensee on or before the buildout deadline. A license can be canceled by a licensee, for example, if it ceases operations and no longer needs the license. Second, FCC could terminate a license before its buildout deadline if the licensee fails to fulfill a condition of the license or violates a rule.[26] Third, the license could expire on or before the buildout deadline. For example, a buildout deadline could be extended past the license expiration date and the licensee could fail to renew the license. The number of licenses in this category varies by service, though a higher percentage of market-based licenses did not reach their buildout deadline compared to site-based licenses.

As discussed above, FCC's enforcement of buildout requirements also involves granting or dismissing extension requests. Across the 5 wireless services we examined, 9 percent of licenses had an extension request. However, this percentage varied substantially across the services (see table 5). Extensions were requested for all of the 39 GHz licenses and half of the 220 MHz phase II licenses. Our analysis of FCC and licensee documents for both services indicated that buildout was largely impeded by lack of available equipment. As shown in table 4 above, both these services also had a relatively high percent of licenses terminated for not meeting buildout requirements. By contrast, less than 1 percent of fixed-microwave and private land-mobile-radio licenses requested extensions. One fixed-microwave licensee we interviewed said that it typically constructs the infrastructure for a new license within 2 to 4 months, so it has not needed to request an extension of the 18-month buildout requirement.

Table 5. Number of Licenses That Did and Did Not Request Extensions by Wireless Service, Data as of September 1, 2013

Service	Number (%) of licenses that did request an extension	Number (%) of licenses that did not request an extension	Total
Broadband PCS	182 (5%)	3,376 (95%)	3,558
220 MHz phase II	482 (50%)	481 (50%)	963
39 GHz	1,672 (100%)	0 (0%)	1,672
Fixed microwave	26 (<1%)	10,452 (100%)	10,478
Private land mobile radio	21 (<1%)	9,525 (100%)	9,546
All market-based services	2,336 (38%)	3,857 (62%)	6,193
All site-based services	47 (<1%)	19,977 (100%)	20,024
Total	2,383 (9%)	23,834 (91%)	26,217

Source: GAO analysis of ULS data.

Notes: See appendix I for additional information on the scope and methodology for the analysis. Percentages may total to more than 100 due to rounding.

We found that FCC granted most extension requests it received for the five wireless services we examined, as shown in table 6. FCC officials said that the Commission seeks to be aggressive but pragmatic in its enforcement of buildout requirements and is flexible on deadlines when it needs to be. FCC officials said that the high rate is due in part to high-quality extension requests. More specifically, they said a licensee typically takes steps before submitting a request—through both internal work and informal discussions with FCC staff—to determine whether it is likely to receive an extension and thus worth the resources to seek an extension. Due to this upfront work, FCC officials said that licensees are likely to submit high-quality extension requests and refrain from submitting unjustified extension requests, which leads to a high percentage of granted requests.

Table 6. Number of Extension Requests Granted and Dismissed by Wireless Service, Data as of September 1, 2013

Service	Number (%) granted	Number (%) dismissed	Total
Broadband PCS	80 (42%)	109 (58%)	189
220 MHz phase II	365 (54%)	315 (46%)	680
39 GHz	1,688 (83%)	345 (17%)	2,033
Fixed microwave	25 (61%)	16 (39%)	41
Private land mobile radio	150 (96%)	6 (4%)	156
All market-based services	2,133 (74%)	769 (27%)	2,902
All site-based services	175 (89%)	22 (11%)	197
Total	2,308 (74%)	791 (26%)	3,099

Source: GAO analysis of ULS data.

Notes: See appendix I for additional information on the scope and methodology for the analysis.

Since a licensee can request more than one extension for a license, the total number of extension requests for each service is greater than the total number of licenses that requested extensions presented in table 5.

Percentages may total to more than 100 due to rounding.

Not surprisingly, we also found that buildout requirements were more likely to be met when all extension requests for a license were granted (see table 7). For Broadband PCS, buildout requirements were met for 84 percent of licenses with granted extension requests while buildout requirements were met for 40 percent of licenses with dismissed extension requests. For 39 GHz, the difference is starker, as buildout requirements were met for 63 percent of licenses with granted extension requests, and buildout requirements were not met for any of the licenses with both granted and dismissed extension requests.[27] Two of the 39 GHz licensees we interviewed said they benefitted from being granted an extension, as the additional time enabled both licensees to meet the buildout requirements for many of their licenses.

Table 7. Outcomes of Licenses by Wireless Service and Extension Request Outcome, Data as of September 1, 2013

Service	Buildout requirement outcome	Request outcome: Number (%) granted	Number (%) dismissed	Number (%) both granted and dismissed[a]	Total
Broadband PCS	Met	63 (84%)	43 (40%)	0 (0%)	106
	Not met or other	12 (16%)	64 (60%)	0 (0%)	76
	Total	75 (100%)	107 (100%)	0 (0%)	182
220 MHz phase II	Met	43 (23%)	2 (2%)	8 (5%)	53
	Not met or other	142 (77%)	128 (99%)	159 (95%)	429
	Total	185 (100%)	130 (100%)	167 (100%)	482
39 GHz	Met	839 (63%)	0 (0%)	0 (0%)	839
	Not met or other	488 (37%)	0 (0%)	345 (100%)	833
	Total	1,327 (100%)	0 (0%)	345 (100%)	1,672
Fixed microwave	Met	10 (100%)	4 (25%)	0 (0%)	14
	Not met or other	0 (0%)	12 (75%)	0 (0%)	12
	Total	10 (100%)	16 (100%)	0 (0%)	26
Private land mobile radio	Met	17 (100%)	1 (33%)	1 (100%)	19
	Not met or other	0 (0%)	2 (67%)	0 (0%)	2
	Total	17 (100%)	3 (100%)	1 (100%)	21

Source: GAO analysis of ULS data.

Notes: See appendix I for additional information on the scope and methodology for the analysis. Percentages may total to more than 100 due to rounding.

[a]Since a licensee can request more than one extension, a license could have requests with different outcomes, that is, both granted and dismissed.

Many experts, licensees, and industry associations we interviewed said that extensions can be beneficial, but some concerns were raised. Some licensees and industry associations we interviewed said that extensions of buildout requirements can provide needed flexibility and be in the public interest. Officials from one industry association said that licensees sometimes encounter an unexpected problem—such as interference with other licensees—and need more time to complete the buildout. Officials from a few industry associations said that extensions provide flexibility when a company has a problem that calls for an extension, especially if a large amount of capital has been invested in the buildout. However, a few licensees and industry associations said that FCC can be inconsistent in granting extensions or that the threshold used to grant extensions was unclear. One licensee and officials from two industry associations, for example, said that FCC has granted many extensions in the past but is now less willing to do so. With respect to the guidance on extensions, officials from one industry association said that

FCC's process provides no certainty as to whether a licensee will get an extension, as they felt that the threshold FCC uses to grant extensions is not clear. FCC officials stated that they review requests for extensions on a case-by-case basis and analyze whether each request meets the legal standard necessary to receive an extension.[28] They added that because the facts are different for each case, an outside party or licensee might observe that the outcomes for extensions were inconsistent, even when the criteria are consistent. Moreover, a few experts had negative opinions on the frequency with which extensions are granted. One expert said that FCC's extensions have set a precedent of extending buildout requirements, which has created an impression that the requirements are not necessarily enforced. Similarly, according to another expert, while the extensions that FCC grants seem reasonable, granting extensions can undermine the purpose of the buildout requirements.

Though infrequent, instances involving FCC delays in reviewing filings—both required notifications and extension requests—can pose problems for licensees. As noted above, we found that 24 220 MHz phase II licenses had filed required notifications that were waiting to be approved or dismissed by FCC. Nineteen of these pending required notifications were filed 4 or more years ago. A few licensees we interviewed said that such delays in processing required notifications for market-based wireless services can create uncertainty. For example, two licensees said that FCC delays in approving a required notification could cause problems or delay the selling or leasing of a license. For extension requests, another licensee said that delays in FCC's response creates risks for licensees and can cause a licensee to expedite construction and spend additional money in case the extension is not granted and the original buildout deadline stands. FCC officials said that they aim to complete reviews as quickly as possible, but do not have a target time for completing reviews. For required notifications, they added that the time needed to complete a review varies depending on the volume of licenses in a service (as many required notifications could be submitted at the same time), the staff resources available, whether a notification contains sufficient information, and what other priorities face FCC or WTB at a given time. For requests for extensions, FCC officials said reviews can take more time compared to other filings as FCC must determine whether the request meets the criteria for an extension and often has to ask a licensee for additional information to better understand the request, among other things. For the 19 filings that have been pending for 4 or more years mentioned above, FCC noted that all these filings aim to demonstrate that the licensees are providing substantial service, and the filings remain pending due to resource limitations, workload priorities, and novel policy, legal, and technical issues the filings present.

Stakeholders Generally Support Buildout Requirements but Identified Changes and Alternative Tools That They Thought Could Better Meet Goals

Most Selected Stakeholders Supported Buildout Requirements but Thought the Requirements More Effectively Met Some Goals Than Others

Nearly all licensees and industry associations we interviewed said that they support FCC's having buildout requirements for wireless services because the requirements help ensure that spectrum will be put to use. In particular, all 10 licensees we spoke with said that they support having buildout requirements for spectrum licenses. Licensees mostly said that the buildout requirements are effective in preventing spectrum warehousing by making licensees accountable for putting the spectrum to use within a specified time frame. For example, one licensee elaborated that there needs to be some kind of buildout requirement in place or the potential for companies to hold spectrum without providing service would increase. Similarly, the majority of industry associations that we spoke with—6 out of 9—support buildout requirements for the same reasons that the licensees cited. However, one association that opposed buildout requirements said the requirements are cumbersome for licensees.

Officials at another industry association were ambivalent, saying that the effectiveness of buildout requirements depends on the type of wireless service, as the requirements make more sense for site-based wireless services than market-based wireless services.

In contrast, spectrum policy experts we spoke with were more mixed in their opinions, with most experts being either ambivalent about or unsupportive of buildout requirements. Two experts who opposed buildout requirements said that there are better alternatives for promoting spectrum efficiency, such as spectrum sharing and encouraging more industry competition. Five experts said that they were ambivalent about the requirements for several reasons, including that the requirements are set too weak or undermined by FCC's granting extensions; as previously mentioned, extensions were requested for 9 percent of licenses we examined, and FCC granted 74 percent of these extension requests. In addition, three experts said that the presence of buildout requirements can lower auction revenues collected by FCC. According to one expert, buildout requirements could force a licensee to deploy a network that might not be the most efficient, which could lower the licensee's expected profits and thus willingness to pay for the license. Only one expert specifically supported buildout requirements without qualification, stating that the requirements make licensees consider whether they will put the wireless license to use before they decide to acquire it.

Beyond these broader stakeholder opinions on buildout requirements, stakeholder opinions on the effectiveness of buildout requirements in meeting commonly cited goals for the requirements were more varied. Of four goals commonly cited for buildout requirements, stakeholders tended to report that buildout requirements are effective in meeting two of these goals: encouraging licensees to provide service in a timely manner and preventing warehousing of spectrum. The stakeholders had mixed views on the effectiveness of buildout requirements in meeting the other two goals: promoting innovative services and promoting services to rural areas.[29]

- *Encouraging licensees to provide service in a timely manner.* Many of the stakeholders whom we interviewed said that the buildout requirements were effective in meeting this goal. More specifically, 9 of 10 licensees and 7 of 9 industry associations said that the requirements were effective in meeting this goal because, for example, they impose construction deadlines that require licensees to put the spectrum to use or surrender the license. In contrast, experts were mixed in their opinions, with 4 of 9 experts saying the requirements were ineffective.[30] For example, one expert said that buildout requirements are not effective in encouraging timely service because FCC does not set buildout requirements that are overly onerous in terms of how long licensees have to meet benchmarks.
- *Preventing the warehousing of spectrum.* Many stakeholders said that the buildout requirements were effective in meeting this goal. In particular, 7 of 10 licensees and 6 of 9 industry associations said that buildout requirements are effective, while experts were mixed. One licensee we interviewed said that buildout requirements can create legitimate pressure for licensees to use the spectrum or offer it for

lease or sale in the secondary market, through which FCC enables licensees to lease or sell portions of the licensed spectrum rights to others. In contrast, 4 of 9 experts said that buildout requirements are ineffective or neither effective nor ineffective in helping FCC meet this goal.[31] For example, one expert said that despite having buildout requirements, FCC's enforcement gives licensees an opportunity to take their time in putting the licensed spectrum to use because the licensees can apply for waivers and extensions and the buildout requirements themselves are not very strict. However, licensees and experts we interviewed generally said they did not believe that spectrum warehousing is a major problem, in their experience. One licensee, for example, said that it does not have an incentive to warehouse spectrum because of high consumer demand for its services.

- *Promoting the provision of innovative services throughout the license areas.* All three groups of stakeholders were mixed in their views on the effectiveness of buildout requirements in promoting innovative services. Licensees and industry associations mostly reported that they thought buildout requirements were neither effective nor ineffective in meeting this goal, and a majority of the experts said that buildout requirements are ineffective in promoting innovative services. For example, three licensees said that innovative services are not directly related to buildout requirements because market forces, such as consumer demand and competition, are what drive innovation. Moreover, two licensees and three experts said that buildout requirements could actually be counter-productive by causing licensees to use older or less innovative technologies to deploy service more quickly.
- *Encouraging the provision of services to rural areas.* Stakeholders were mixed in their views about whether buildout requirements help promote services in rural areas. For example, four licensees said that the buildout requirements were effective while four said they were ineffective, and five experts said that the requirements were effective while two experts said they were ineffective. The licensees and industry association representatives noted that building out to rural areas is difficult and expensive and the high costs associated with construction in these areas are rarely recovered by providing service to sparsely populated areas with few customers. A few stakeholders across all three groups added that geographic coverage requirements

are more effective in promoting rural service than population coverage requirements. For example, one industry association said that geographic coverage requirements better promote rural buildout, particularly for licenses covering large geographic areas; if a large geographic-area license has a population coverage requirement, a licensee might be able to meet the requirement by serving the relatively densely populated areas and leaving the rural areas unserved.

Stakeholders Identified a Variety of Changes and Alternative Tools That They Thought Could Better Meet Commonly Cited Goals for Buildout Requirements

Changes to Buildout Requirements

While buildout requirements are generally supported, some stakeholders we spoke with said that the requirements were not effective in meeting some of the commonly cited goals, in particular promoting innovative services and services to rural areas, as discussed above. Therefore, 22 of 28 stakeholders we spoke with identified changes that they said could improve the effectiveness of buildout requirements for wireless services to meet the goals identified above, in particular for market-based services.[32] The most frequently mentioned changes or enhancements include the following:

- *More clarity.* Four licensees, two industry associations, and three experts said additional clarity could make buildout requirements more effective. The four licensees, for example, reported that more clarity in the service rules could allow both FCC and licensees to better meet goals by removing uncertainty. Specifically, stakeholders said more clarity could be provided through greater detail about what could constitute substantial service[33] or about the engineering parameters licensees should use in their required notifications. According to one licensee, any clarification on a required process or rule is helpful, and for buildout requirements, more specific guidance might help eliminate some back and forth needed for FCC to approve a required notification. FCC officials said that the Commission sets specific requirements for waivers and extension requests, as well as specific buildout requirements, and that it reviews licensee notifications and requests on a case-by-case basis.

- *More robust and transparent enforcement.* Three industry associations, two experts, and a licensee said that the self-certification process, while efficient and appropriate, could be bolstered by more visible enforcement, such as using spot checks to verify licensees' required notifications. For example, one expert said that to ensure effectiveness of the buildout requirements, FCC could conduct random spot checks to see that licensees are providing services upon meeting the buildout requirement. Similarly, to increase transparency, an industry association said that FCC could better educate licensees about the administrative aspects of filing requests or notifications and then conduct spot checks and issue fines if licensees are not providing services. These licensees and industry associations also said that more consistent enforcement entails more transparency and consistency in FCC's processes for granting extensions and waivers.
- *Different penalties.* Some stakeholders—two licensees and two experts—said that FCC could change the penalty for not meeting a buildout requirement. Many licensees, industry associations, and experts said that the penalty of license termination was too strict. Specifically, officials from one industry association said that with termination, licensees face the loss of all their investment in constructing infrastructure if they have to surrender the license for not meeting buildout requirements. Some of these stakeholders favored a use-it-or-share-it approach whereby a licensee would have to make spectrum for which it did not meet a buildout requirement available to others through leasing or sharing. According to one expert, use it or share it could create opportunities for other users to benefit by using spectrum that is lying fallow more immediately, even if only temporarily, or providing a stronger incentive for a licensee to make secondary market arrangements to put spectrum to use.
- *More opportunities to align licensees' goals with buildout requirements.* Two licensees and one industry association noted that FCC's buildout requirements do not necessarily align with a licensee's business plans, particularly for market-based services. Two licensees said FCC could provide more upfront feedback to licensees on whether a licensee's plan to meet a buildout requirement would be accepted by the Commission. These licensees said that through such early interaction with FCC, licensees might invest in building their systems and meet buildout requirements in tandem, rather than potentially having to invest in additional infrastructure simply to meet

a buildout requirement to save a license. For example, one licensee said it currently has to consider two parallel tracks—its business plan and FCC buildout requirements—when building a system, a situation that can increase costs and make buildout less efficient. The licensee believed it could more efficiently construct its system if these two tracks could be brought closer together early in the license term. FCC officials told us that licensees can engage with FCC to obtain informal guidance before filing notifications to discuss the sufficiency of their plans and avoid potential problems. One industry association also said that FCC could provide licensees with additional ways to demonstrate meeting a buildout requirement, beyond a single requirement or safe harbor for substantial service, to distinguish licensees that are warehousing spectrum from those who are working to put it to use.

Alternative Tools

In addition to changes to buildout requirements, stakeholders from each of the three groups we spoke with identified alternatives to buildout requirements that they said could better meet the four commonly cited goals, including provision of innovative services and service to rural areas. We also identified additional support for stakeholder-identified tools and other tools through a review of our previous reports on spectrum management and comments filed in response to FCC proceedings related to buildout requirements for spectrum licenses.[34] Some alternative tools could be used in place of buildout requirements, and others could complement the buildout requirements with the intent to better meet the goals and promote efficient use of spectrum. Table 8 summarizes the alternatives to buildout requirements that stakeholders identified as tools that could better meet each of the four commonly cited goals.

- *Secondary markets.* FCC enables licensees to make transactions through secondary markets, such as leasing spectrum rights to other licensees; this process facilitates licensees selling or leasing unneeded spectrum rights by negotiating their own terms.[35] Three licensees and two experts said that these transactions could promote the provision of timely service by allowing for accelerated transactions to a licensee that wants to deploy wireless services sooner and without the additional time needed for FCC review of the transaction. Some of these licensees and experts also said that this alternative may help better meet the goal of preventing spectrum warehousing by allowing a licensee, which does not want to deploy service in the spectrum in the near future, to recover costs by leasing or selling the spectrum rights to others that want to put the spectrum to use more immediately.

Table 8. Stakeholder-Identified Replacements for and Complements to FCC Buildout Requirements

Goal	Secondary markets	Reliance on market forces	Flexible use licenses	Spectrum sharing	Smaller License areas	Subsidies	Spectrum usage fees
Encouraging licensees to provide timely service	X	X	X	X			
Promoting innovative services		X	X				
Providing service to rural areas					X	X	
Preventing warehousing of spectrum	X			X			X

Source: GAO analysis of interviews and reports.

- *Reliance on market forces.* A few licensees and experts said that relying more on market forces could help spur competition and ultimately encourage licensees to provide timely and innovative services. FCC already relies on market forces to some degree by, for example, auctioning licenses. Three licensees and two experts we spoke with said that FCC could further implement or bolster policies to promote competition that better motivate licensees to build or expand their networks and provide services more quickly than buildout requirements alone. Furthermore, one expert said that FCC already promotes competition among existing licensees as well as encouraging entry for others through the auctioning process, so buildout requirements are not needed in settings where there is sufficient competition to encourage licensees to acquire and put spectrum to use.
- *Flexible-use licenses.* With traditional licenses, the use or service is limited to the specific terms of the license (e.g., broadcast a television signal in a specific geographic market), but flexible-use licenses allow

for a wider array of uses without having to seek additional FCC authorization.[36] Therefore, several stakeholders we spoke with said that FCC could do more to allow flexible-use licenses and that this might speed up wireless service deployment and help meet the goal of promoting timely service. FCC officials said that the Commission does propose to issue flexible-use licenses when circumstances permit but that flexible-use licenses are not appropriate for allowing certain services in specific bands, such as broadcast services in a mobile wireless band, or when technical limitations of a band limits flexibility. According to stakeholders, flexible-use licenses could also help promote innovative services. One licensee, for example, said that flexible-use licenses allow it to update its networks and technology without changing bands or asking FCC to modify licenses.

- *Spectrum sharing.* Through our interviews and review of two FCC proceedings, three experts and one licensee reported that enabling spectrum sharing may encourage licensees to put licensed spectrum to use while allowing them to increase efficiency in their business plans. This cooperative use of spectrum, with multiple users agreeing to access the same spectrum at different times or locations, could allow licensees to provide service more quickly and help prevent warehousing of spectrum.[37] For example, one licensee said that having a spectrum-sharing policy would be a good supplement to buildout requirements; specifically, if a licensee does not meet its buildout requirements, FCC could require that the licensee negotiate sharing or leasing for the unused part of the license to help put it to use in timely manner. Similarly, one expert said sharing could enable others to put spectrum to use in cases where the licensee is not ready to use the spectrum, thus putting the spectrum to use more quickly, and might help discourage licensees from warehousing spectrum.
- *Smaller license areas.* Through our interviews and review of FCC proceedings, two licensees and two experts we spoke with said smaller geographic area licenses could better encourage service to rural areas. One licensee said that for market-based services, auctioning smaller-sized licenses could allow entities, such as rural wireless licensees, to bid on the specific areas they want to serve. One industry association commented that a larger inventory of smaller, and likely more affordable, licenses might attract the small and rural providers that best know and can best serve rural areas. Also, as previously discussed, another industry association commented that

licensing exclusively by larger blocks could disfavor competition and discourage deployment of services in rural and less densely populated areas.

- *Subsidies.* Through interviews, two licensees and two experts said that using subsidies would be a more effective way to help promote service to rural areas. One example of a subsidy is the Universal Service Fund, in which FCC establishes programmatic goals and distributes a subsidy to assist in efforts to increase nationwide access to advanced wireless services.[38] Two experts said that the buildout requirements are vague and represent a crude and untargeted approach to addressing a specific goal like promoting services to underserved and unserved areas in rural locations. For example, one expert said that while a specially designed buildout requirement may be effective in prompting a licensee to provide service to rural areas, using a subsidy to procure an entity that is willing to provide service would be a more direct and effective way to meet the goal. One licensee said that buildout requirements have a limited ability to promote service to rural areas since it is often not economical for a licensee to build a system in areas that are sparsely populated; in these cases, subsidies can better encourage licensees to serve these areas.
- *Spectrum usage fees.* A few licensees we interviewed said spectrum usage fees may be a good alternative to buildout requirements to help prevent spectrum warehousing, particularly for licenses not obtained through auctions. Usage fees could help encourage licensees to use spectrum more efficiently or pursue sharing opportunities once they bear the opportunity cost of letting licensed spectrum sit idle.[39] For example, one licensee said that if spectrum is made available for free, a licensee may have less incentive to put it to use or use it efficiently compared to a licensee that bought its spectrum at auction. For this reason, a few licensees noted that when FCC began using auctions to assign spectrum, there was a debate about whether buildout requirements were needed for auctioned licenses since those who paid to acquire spectrum have demonstrated their commitment to use the spectrum by paying for it.

AGENCY COMMENTS

We provided a draft of this report to FCC for review and comment. FCC provided technical comments that we incorporated throughout the report as appropriate.

Mark L. Goldstein
Director,
Physical Infrastructure

APPENDIX I: OBJECTIVES, SCOPE, AND METHODOLOGY

This report examines Federal Communications Commission (FCC) buildout requirements for wireless services and the efficient use of spectrum. In particular, this report provides information on (1) the buildout requirements established by FCC for spectrum licenses for wireless services, (2) the extent that FCC follows its process to enforce buildout requirements for wireless services, and (3) stakeholder opinions on the extent that goals for buildout requirements have been met.

To describe FCC buildout requirements for wireless services, we reviewed FCC regulations and guidance on buildout requirements for services that use spectrum. We also interviewed FCC officials to understand which services have buildout requirements, the general process used to set buildout requirements for a service, and what factors FCC considers when setting buildout requirements for a service. According to FCC, the Wireless Telecommunications Bureau (WTB) is responsible for granting and monitoring licenses for wireless services that use spectrum. For this review, we focused on guidance and processes related to buildout requirements for licenses for wireless services, which amounts to nearly 2 million licenses.[1]

To describe FCC's enforcement process, we reviewed FCC regulations and guidance to determine the steps FCC takes to monitor and enforce buildout requirements for wireless services. In addition, we interviewed FCC officials to learn about different parts of the enforcement process to determine licensee responsibilities and actions, FCC responsibilities and actions, and which FCC actions are automated in the Universal Licensing System (ULS) licensing database.

To examine FCC's enforcement of buildout requirements for wireless services, we selected five wireless services and analyzed data on them from FCC's ULS database. We selected the five services to ensure variety in type of service or use, type of buildout requirement (e.g., population coverage or substantial service), how licenses were assigned (e.g., auctions), and the number of licenses in the service. We also considered recommendations from FCC officials and other interviewees when selecting among wireless services. As a result, we selected three market-based services—broadband personal communications service (PCS), 220 megahertz (MHz) phase II, and 39 gigahertz (GHz)—and two site-based services—industrial/business private land mobile radio below 700 MHz and fixed microwave.[2] Tables 1 and 3 in the report provide information on how each selected service aligns with the criteria we used to select wireless services. The results of the data analysis for these five services are not generalizable to other wireless services.

For the five selected wireless services, we analyzed data for licenses that would have buildout requirements on or before December 31, 2012.[3] We picked December 31, 2012, to allow sufficient time after the buildout deadline for any licensee or FCC action—such as FCC review of a licensee's notification that it met the requirement—to occur and be entered in ULS. For the market-based services, we included all licenses that would have a buildout requirement on or before December 31, 2012, based on the auction dates for the licenses and the length of the buildout requirement in regulation. For example, we included Broadband PCS licenses awarded at auctions between 1995 and 2007; since these licenses have a 5-year buildout requirement, the buildout deadlines fell between 2000 and 2012.[4] For site-based services, we similarly sought to include licenses that would have buildout requirements due on or before December 31, 2012. However, due to the high volume of licenses in these two services, we limited our analysis to new licenses that would have a buildout requirement during calendar year 2012; that is, we did not include modifications to existing licenses, for which FCC also includes buildout requirements. Private land-mobile-radio licenses below 700 MHz, for example, have a 12-month buildout requirement, so we included new licenses granted during calendar year 2011 that would have a buildout deadline during calendar year 2012. Appendix III contains additional information on the number and type of licenses in each wireless service included in our analysis, such as the auction numbers and dates for market-based services.

For each service, we analyzed license data to determine the outcomes of buildout requirements and examine FCC's enforcement of buildout requirements. We used license and application data from the ULS public access downloads as of September 1, 2013.[5] In particular, we examined (1) the number of licenses that did and did not have buildout requirements; (2) the outcomes for licenses that had buildout requirements; (3) the number of licenses with requests for extensions; and (4) for licenses with requests, whether the request was granted or dismissed.

With respect to the outcomes for licenses that had buildout requirements, we examined the number of licenses in each wireless service that:

- met the requirement, including whether it met the requirement on time or late;[6]
- did not meet the requirement and was terminated;
- did not meet the requirement and remained active after the buildout deadline; and
- did not reach the buildout deadline, meaning that the license was (1) canceled on or before the buildout deadline, (2) otherwise terminated before the buildout deadline, or (3) expired on or before the buildout deadline.

For licenses with outcomes that did not appear to align with FCC's enforcement process, we conducted additional research to understand the circumstances for these licenses. Specifically, (1) licenses that met the buildout requirement late and (2) licenses that did not meet the requirement and were not terminated on the buildout deadline. For these licenses, we reviewed additional information using ULS's online license search to determine whether FCC followed its enforcement processes. We also asked FCC officials about the general circumstances surrounding these licenses. Overall, a small percentage of licenses had one of these outcomes.

Since some licenses had more than one buildout requirement and thus could have more than one outcome, we developed rules to classify these licenses. For market-based licenses with more than one buildout requirement on or before December 31, 2012, we generally classified the license's buildout outcome by the outcome for the second buildout requirement. For example, if a first buildout requirement was met but the license was canceled before the second buildout requirement, we classified the outcome as "canceled on or before buildout requirement." However, to be classified as "met," both the first and second requirement had to be met for a license. For the site-based services, a license can authorize multiple frequencies, and each frequency could have a buildout requirement. For each site-based license, we assessed the outcomes for all frequencies and used this information to report an outcome for the license. For example, if the buildout requirements were met for all frequencies, we classified the outcome as "met." If the buildout requirements were met for some frequencies but not met for other frequencies (meaning that those frequencies were terminated), we classified the license as "some met/some not met." When examining extension requests for licenses in all five services, we assessed whether any extension request was filed for the license.

Based on interviews with FCC officials, as well as reviewing system documentation and electronic data testing, we determined that these data were sufficiently reliable for our purposes. Appendix III contains detailed results of the analysis of ULS data for each of the five selected wireless services.

To assess the effectiveness of FCC's enforcement, we also selected and interviewed a sample of industry associations and licensees. We conducted semi-structured interviews with industry associations and licensees to gather their opinions on FCC's enforcement process, including the clarity of FCC guidance on buildout requirements, the timeliness of FCC responses to licensee requests and other applications, and their experiences with ULS. We selected both industry associations and licensees to cover the five selected wireless services, and we further selected licensees to ensure variety in licensee type and size. The opinions from these industry associations and licensees are not generalizable.

To gather stakeholder opinions on the effectiveness of buildout requirements, we selected and interviewed a sample of industry stakeholders, including spectrum policy experts, industry associations, and licensees. To determine the goals for buildout requirements, we reviewed statute, FCC documents from recent rulemakings, and other FCC budget and performance documents to identify frequently cited goals for buildout requirements. In addition to the general goal of promoting efficient or productive use of spectrum, we identified the following four goals:

- encouraging licensees to provide service in a timely manner,
- promoting the provision of innovative services throughout the license areas,
- encouraging provision of services to rural areas, and
- preventing the warehousing of spectrum.

We conducted semi-structured interviews with stakeholders to gather their opinions on the extent that buildout requirements meet each of the four goals as well as reasons or examples to support these opinions (see table 9). We also asked what changes, if any, could make buildout requirements more effective and what alternative tools FCC could use to more directly or better meet the four goals. To select experts, we included individuals based on participation in recent GAO reviews on spectrum policy, publications on spectrum policy, and recommendations from other interviewees.[7] In particular, we sought to interview individuals who appeared at least twice across the criteria or participated in at least two recent GAO reviews. We selected industry associations and licensees as described above. The views of the selected stakeholders we interviewed are not generalizable.

Table 9. Organizations and Individuals Interviewed

Industry associations
Central Station Alarm Association
Competitive Carriers Association
Enterprise Wireless Alliance
Fixed Wireless Communications Coalition
Forest Industries Telecommunications
Land Mobile Communications Coalition
PCIA – The Wireless Infrastructure Association
Telecommunications Industry Association
Utilities Telecom Council
Experts
Coleman D. Bazelon, The Brattle Group
Michael Calabrese, New American Foundation
Peter Cramton, University of Maryland
Donald Evans, Fletcher, Heald & Hildreth, PLC
Harold Feld, Public Knowledge
Thomas W. Hazlett, George Mason University School of Law
William Lehr, Massachusetts Institute of Technology
Gregory L. Rosston, Stanford Institute for Economic Policy Research
Scott Wallsten, Technology Policy Institute
Licensees
CSpire
Day Wireless Systems
IDT Spectrum
National Rural Telecommunications Cooperative
North Dakota Network Co.
Peñasco Valley Telephone Cooperative, Inc.

Southern Company
Sprint
T-Mobile
Verizon

Source: GAO.

To supplement these interviews, we reviewed our previous reports on spectrum management and filings in two recent FCC proceedings that sought comments on buildout requirements for wireless services. We identified and reviewed filings in the dockets for the following two proceedings: 07-293, *In the Matter of Establishment of rules and policies for the Digital Audio Radio Satellite Service in the 2310-2360 MHz Frequency Band*[8] and 12-268, *In the Matter of Expanding the Economic and Innovation Opportunities of Spectrum Through Incentive Auctions.*[9] For each proceeding, we reviewed filings made by licensees, industry associations, and other companies and associations and summarized opinions on whether buildout requirements are effective and any changes that could be made to improve buildout requirements.

APPENDIX II: BUILDOUT REQUIREMENTS FOR SPECTRUM LICENSES

Buildout Requirements for Services Licensed by the Wireless Telecommunications Bureau and the Public Safety and Homeland Security Bureau

Table 10. Buildout Requirements for Private and Commercial Land-Mobile Radio Services, as of January 2014

Radio service/frequency band	Construction/coverage requirements	Radio service code	47 C.F.R. Section
220 MHz (Public Safety)	Construct and place into operation within 12 months of initial license grant date, or the authorization cancels automatically and must be returned to the Commission. Alternatively, may apply for extended implementation - up to 3 years may be authorized for constructing and placing a system in operation.	QM	90.725(f), 90.727

Table 10. (Continued)

Radio service/frequency band	Construction/coverage requirements	Radio service code	47 C.F.R. Section
220 MHz Phase I (Nationwide Licenses)	Construct and place into operation at least 10 percent of designated geographic areas within 2 years of initial license grant, 40 percent within 4 years, 70 percent within 6 years, and 100 percent of the designated geographic areas within 10 years of initial license grant. Each benchmark has a requirement for base stations to be operational in specific urban areas.	NC	90.725(a)
220 MHz Phase I (Non-Nationwide Licenses)	Construct and place into operation within 12 months of initial license grant date.	QD, QO, QT	90.725(f), 90.757(b)
220 MHz Phase II (Nationwide Licenses)	Provide coverage to a composite area of 750,000 square kilometers or 37.5 percent of the US population within 5 years of initial license grant, and 1,500,000 square kilometers or 75 percent of the population within 10 years of initial license grant. Alternatively, nationwide licensees may meet the construction requirements by demonstrating an appropriate level of substantial service at the 5- and 10-year benchmarks.	QA	90.769
220 MHz Phase II (Economic Area or Regional Licenses)	Provide coverage to at least 1/3 of the population in the licensed area within 5 years of initial license grant, and 2/3 of the population within 10 years of initial license grant. Alternatively, Economic Area or Regional licensees may meet their construction requirements by providing an appropriate level of substantial service at 5- and 10- year benchmarks.	QA	90.767
Below 700 MHz (except 220 MHz)	Must be placed into operation within 12 months of license grant, or the authorization cancels automatically and must be returned to the Commission. Alternatively, a local government in the Public Safety Pool may apply for extended implementation - up to 5 years may be authorized for constructing and placing a system in operation.	IG, IK, YG, YK, PW, YW, RS	90.155, 90.629

Radio service/frequency band	Construction/coverage requirements	Radio service code	47 C.F.R. Section
700 MHz Narrowband Public Safety (except State Licenses)	Construct and place into operation within 12 months of initial grant. Alternatively, may apply for extended implementation - up to 5 years may be authorized for constructing and placing a system in operation.	SG, SY	90.551, 90.155(b)
700 MHz Narrowband Public Safety State Licenses	State must certify substantial service to 1/3 of its population by June 13, 2014, and 2/3 of its population by June 13, 2019. Licensee must certify that the radio system has been approved and funded for implementation by the deadline date.	SL	90.529
800 and 900 MHz Trunked (Excluding Specialized Mobile Radio)	Construct and place into operation within 1 year of initial license grant. Alternatively, may apply for extended implementation - up to 5 years may be authorized for constructing and placing a system in operation.	YB, YE, YF, YI, YJ, YO, YP, YU	90.631(e), 90.629
800 and 900 MHz Conventional (Excluding Specialized Mobile Radio)	Construct and place into operation within 1 year of initial license grant. Alternatively, may apply for extended implementation - up to 5 years may be authorized for constructing and placing a system in operation.	GB, GE, GF, GI, GJ, GO, GP, GU	90.633(c), 90.629
800 and 900 MHz Trunked (Specialized Mobile Radio)	Construct and place into operation within 1 year of license grant.	YX, YS, YM, YL	90.631(e-f)
800 and 900 MHz Conventional (Specialized Mobile Radio)	Construct and place into operation within 1 year of license grant.	GX, GR, GM, GL	90.633(c)
800 MHz Geographic (Economic Area) (Channel Blocks D through V)	Provide coverage to at least 1/3 of the population in the licensed area within 3 years of initial license grant, and 2/3 of the population within 5 years of initial license grant. Alternatively, licensees may provide substantial service to the licensed area within 5 years of initial license grant.	YC, YH	90.685(b)
800 MHz Geographic (Economic Area) (Channel Blocks A, B, C)	Provide coverage to at least 1/3 of the population in the licensed area within 3 years of initial license grant, and 2/3 of the population within 5 years of initial license grant.	YC, YH	90.685(b)

Table 10. (Continued)

Radio service/frequency band	Construction/coverage requirements	Radio service code	47 C.F.R. Section
	In addition, licensees must construct 50 percent of the total number of channels in their spectrum block in at least one location in their licensed area within 3 years of initial grant and retain such usage for the remainder of the construction period. Alternatively, licensees may provide substantial service to the licensed area within 5 years of initial license grant. In addition, licensees must construct 50 percent of the total number of channels in their spectrum block in at least one location in their licensed area within 3 years of initial grant and retain such usage for the remainder of the construction period.		
900 MHz Geographic (Major Trading Area)	Provide coverage to at least 1/3 of the population in the licensed area within 3 years of initial license grant, and 2/3 of the population within 5 years of initial license grant. Alternatively, licensees may provide substantial service to their licensed area within 5 years of initial license grant. If substantial service at 5 years is elected, licensees must notify FCC within 3 years that it is electing the substantial service option at 5 years.	YD	90.665(c)
4940-4990 MHz (Public Safety 4.9 GHz)	Fixed point-to-point stations must be placed into operation within 18 months of initial license grant, or the authorization for that station cancels automatically. There is no time limit for which base and temporary fixed stations authorized under a 4940–4990 MHz band license must be placed in operation.	PA	90.1209

Radio service/frequency band	Construction/coverage requirements	Radio service code	47 C.F.R. Section
Intelligent Transportation Service, Dedicated Short Range Communications Service Roadside Units	Construct and place into operation within 12 months from the date of registration. Licensees must notify the Commission when registered units are placed in operation within their construction period.	IQ (Public-Safety), QQ (Non-Public Safety)	90.155(i)

Source: FCC.

Table 11. Buildout Requirements for Location and Monitoring Services, as of January 2014

Radio service/frequency band	Construction/coverage requirements	Radio service code	47 C.F.R.Section
Geographic Licenses (Economic Area)	Multilateration location service available to 1/3 of the population within 5 years of initial license grant and 2/3 of the population within 10 years of initial license grant. Alternatively, licensees may provide substantial service to their licensed area within the appropriate 5- and 10-year benchmarks.	LS	90.155(d)
Site-Based Licenses	Construct and place into operation within 12 months of license grant.	LN, LW	90.155(a)

Source: FCC.

Table 12. Buildout Requirements for Cellular Services, as of January 2014

Radio service/frequency band	Construction/coverage requirements	Radio service code	47 C.F.R. Section
First system on each channel block, markets 1-90	At least partially constructed and begin providing service to subscribers within 36 months of initial license grant.	CL	22.946(a), 22.947
First system on each channel block, all markets except for markets 1-90 and Gulf of Mexico Exclusive Zone.	At least partially constructed and begin providing service to subscribers within 18 months of initial license grant.	CL	22.946(a), 22.947

Table 12. (Continued)

Radio service/frequency band	Construction/coverage requirements	Radio service code	47 C.F.R. Section
Unserved Areas or Specific Facilities	Construct and place into operation within 12 months.	CL	22.946(c), 22.947

Source: FCC.

Table 13. Buildout Requirements for Broadband Personal Communications Service (PCS), as of January 2014

Radio service/frequency band	Construction/coverage requirements	Radio service code	47 C.F.R.Section
30 MHz Licenses	Service to at least 1/3 of the population in licensed area within 5 years of initial license grant, and 2/3 of the population within 10 years of initial license grant. Alternatively, licensees may provide substantial service to their licensed area within the appropriate 5- and 10-year benchmarks.	CW	24.203(a)
10/15 MHz Licenses	Service to at least 1/4 of the population in licensed area within 5 years of initial license grant, or alternatively substantial service within 5 years of initial license grant.	CW	24.203(b)

Source: FCC.

Table 14. Buildout Requirements for Narrowband Personal Communications Service (PCS), as of January 2014

Radio service/frequency band	Construction/coverage requirements	Radio service code	47 C.F.R. Section
Nationwide Licenses	Provide coverage to a composite area of 750,000 square kilometers or 37.5 percent of the US population within 5 years of initial license grant, and 1,500,000 square kilometers or 75 percent of the population within 10 years of initial license grant. As an alternative to the construction requirements listed above, all narrowband PCS licensees may provide substantial service to their licensed area within 10 years of initial license grant.	CN	24.103(a),(d)

Radio service/frequency band	Construction/coverage requirements	Radio service code	47 C.F.R. Section
	Licensees electing substantial service must notify FCC at the 5 year benchmark that they plan to satisfy the substantial service requirement.		
Regional Licenses	Provide coverage to a composite area of 150,000 square kilometers or 37.5 percent of the population of the service area within 5 years of initial license grant, and 300,000 square kilometers or 75 percent of the population within 10 years of initial license grant. As an alternative to the construction requirements listed above, all narrowband PCS licensees may provide substantial service to their licensed area within 10 years of initial license grant. Licensees electing substantial service must notify FCC at the 5 year benchmark that they plan to satisfy the substantial service requirement.	CN	24.103(b), (d)
Major Trading Area Licenses	Provide coverage to a composite area of 75,000 square kilometers or 25 percent of geographic area, or 37.5 percent of the population in licensed area within 5 years of initial license grant; and 150,000 square kilometers or 50 percent of the geographic area, or 75 percent of the population within 10 years of initial license grant. As an alternative to the construction requirements listed above, all narrowband PCS licensees may provide substantial service to their licensed area within 10 years of initial license grant. Licensees electing substantial service must notify FCC at the 5 year benchmark that they plan to satisfy the substantial service requirement.	CN	24.103(c), (d)
1910-1915/1990-1995 MHz PCS	Substantial service showing within 10 years of the date of initial issuance or renewal.	CY	24.203(d)

Source: FCC.

Table 15. Buildout Requirements for Paging and Radiotelephone Services, as of January 2014

Radio service/frequency band	Construction/coverage requirements	Radio service code	47 C.F.R. Section
Site-Based Licenses - Part 22	Construct within 1 year of license grant.	CA, CB, CD, CG, CO, CR	22.511
Site-Based Licenses - Part 90, 929-930 MHz, Exclusive Channels	Construct within 1 year of license grant.	GS	90.493(b), 22.511
Site-Based - Part 90, 929-930 MHz, Shared Channels	Construct and place into operation within 12 months of license grant.	GS	90.155(a)
Geographic Area Commercial Aviation Air-Ground Systems	Licensees authorized to use more than one megahertz (1 MHz) of the 800 MHz commercial aviation air-ground spectrum allocation must make a showing of substantial service within 5 years of grant of authorization. Substantial service may be demonstrated, but is not limited to, either of the following "safe harbor" provisions: (1) Construction and operation of 20 ground stations with at least one ground station located in each of the 10 Federal Aviation Administration regions; or, (2) Provision of service to the airspace of 25 of the 50 busiest airports.	CJ	22.873
Geographic Area Paging Licenses	Provide coverage to 1/3 of the population within the licensed area within 3 years of initial license grant, and 2/3 of the population within 5 years of initial license grant. Alternatively, licensees may provide substantial service to the licensed area within 5 years of initial license grant. Licensees electing the substantial service option must notify FCC of their intention to do so within their 3-year construction benchmark.	CP, CZ, GC	22.503(k)

Source: FCC.

Table 16. Buildout Requirements for Miscellaneous Wireless Communication Services, as of January 2014

Radio service/frequency band	Construction/coverage requirements	Radio service code	47 C.F.R. Section
Lower 700 MHz (Block A 698-704 and 728-734 MHz; Block B 704-710 and 734-740 MHz; Block E 722-728 MHz)	Provide signal coverage and offer service for at least 35 percent of the geographic area of each of its license authorizations no later than June 13, 2013 (or within four years of initial license grant if later than June 13, 2009). If licensee fails to do so, the license authorization will be reduced by 2 years and licensee may lose authority to operate in remaining unserved areas and other penalties may apply. Provide service over at least 70 percent of the geographic area of authorization by the end of the license term. If licensee fails to do so, licensee's authorization for unserved areas will terminate automatically and may be subject to license termination and other penalties. If results of first auction in which licensees for such authorizations in Blocks A, B, and E are offered do not satisfy the reserve price for the applicable block, licensee must provide signal coverage and offer service over at least 40 percent of the population no later than June 13, 2013 (or within 4 years of initial license grant if later than June 13, 2009). If licensee fails to do so, the license authorization will be reduced by two years and licensee may lose authority to operate in remaining unserved areas and other penalties may apply. Provide signal coverage and offer service to at least 75 percent of the population by the end of the license term. If licensee fails to do so, licensee's authorization for unserved geographic portions will terminate automatically and licensee may be subject to license termination and other penalties.	WY	27.13, 27.14(g)(1), 27.14(g)(2), 27.14(i)

Table 16. (Continued)

Radio service/frequency band	Construction/coverage requirements	Radio service code	47 C.F.R. Section
700 MHz Lower Band (Block C 710-716, 740-746 MHz and Block D 716-722 MHz)	Provide substantial service to its licensed area within 10 years of initial license grant.	WZ	27.13, 27.14(a)
700 MHz Upper Band (Block C 746-757 MHz and 776-787 MHz)	Provide signal coverage and offer service over at least 40 percent of the population no later than June 13, 2013 (or within four years of initial license grant if later than June 13, 2009). If licensee fails to do so, the license authorization will be reduced by 2 years and licensee may lose authority to operate in remaining unserved areas and other penalties may apply. Provide signal coverage and offer service to at least 75 percent of the population by the end of the license term. If licensee fails to do so, licensee's authorization for unserved geographic portions will terminate automatically and licensee may be subject to license termination and other penalties.	WU	27.14(h)
700 MHz Upper Band (Block D 758-763, 788-793 MHz)	Provide signal coverage and offer service over at least 75 percent of the nationwide Upper D Block population within 4 years from June 13, 2009, 95 percent of the population of the nationwide license area within 7 years, and 99.3 percent of the population of the nationwide license area within ten years.	WP	27.14(m)
700 MHz Guard Bands	Provide substantial service to its licensed area within 10 years of initial license grant. Submit an annual report providing information about the manner in which its spectrum is being utilized.	WX	27.13, 27.14(a), 27.607
1.3 GHz (1390-1392 MHz), Market Area	Provide substantial service to its licensed area within 10 years of initial license grant.	BA	27.13, 27.14(a)
1.4 GHz (1392-1395 MHz and 1432-1434 MHz), Market Area	Provide substantial service to its licensed area within 10 years of initial license grant.	BB	27.13, 27.14(a)

Radio service/frequency band	Construction/coverage requirements	Radio service code	47 C.F.R. Section
1.6 GHz (1670-1675 MHz), Market Area	Provide substantial service to its licensed area within 10 years of initial license grant.	BC	27.13, 27.14(a)
Advanced Wireless Service (AWS) (1710-1755 MHz and 2110-2155 MHz)	Provide substantial service to its licensed area within the prescribed license term. Authorizations will have a term not to exceed 10 years from the initial license grant, except that authorizations issued on or before December 31, 2009, will have a term of 15 years.	AW	27.13, 27.14
AWS-4 (2000-2020 MHz and 2180-2200 MHz)	Within 4 years from the date of license, provide reliable signal coverage and offer service to at least 40 percent of its total AWS-4 population. Within 7 years of the date of license, provide reliable signal coverage and offer service to at least 70 percent of the population in each of its license areas.	AD	27.13, 27.14
2.3 GHz Wireless Communications Services (WCS) (2305-2320 MHz and 2345-2360 MHz)	Licensees operating mobile or point-to-multipoint systems must provide reliable signal coverage and offer service to at least 40 percent of the license area's population within 48 months and to at least 75 percent of the license area's population within 78 months. If a licensee demonstrates that 25 percent or more of the license area's population is within a zone requiring coordination with an aeronautical mobile telemetry facility or with the National Aeronautics and Space Administration's Deep Space Network facility in Goldstone, California, the population benchmarks are reduced to 25 and 50 percent, respectively. Licensees operating point-to-point fixed systems, except those deployed in the Gulf of Mexico license area, must construct and operate a minimum of 15 point-to-point links per million persons (one link per 67,000 persons) in a license area within 48 months, and 30 point-to-point links per million persons (one link per 33,500 persons) in a licensed area within 78 months.	WS	27.13, 27.14(p)

Table 16. (Continued)

Radio service/frequency band	Construction/coverage requirements	Radio service code	47 C.F.R. Section
	Licensees operating point-to-point fixed systems in the Gulf of Mexico license area must demonstrate that they have constructed and are operating a minimum of 15 point-to-point links at 48 months and 78 months.		
Broadband Radio Service	Provide substantial service to its licensed area by May 1, 2011.	BR	27.14
Educational Broadband Service	Provide substantial service to its licensed area by November 1, 2011.	ED	27.14

Source: FCC.

Table 17. Buildout Requirements for Maritime Public Coast Services, as of January 2014

Radio service/frequency band	Construction/coverage requirements	Radio service code	47 C.F.R. Section
Very High Frequency (VHF), Geographic Licenses	Provide substantial service to its licensed area within 5 years of initial license grant and again within 10 years.	PC	80.49(a)(1)
VHF, Site-Based Licenses	Construct and place into operation within 12 months from date of grant.	MC	80.49(a)(1)
Low Frequency, Medium Frequency, High Frequency Site-Based Licenses	Construct and place into operation within 12 months from date of grant.	MC	80.49(a)(2)
Automated Maritime Telecommunications System (AMTS), Site-Based	Construct and place into operation within 2 years from date of grant.	MC	80.49(a)(3)
AMTS, Geographic Licenses	Provide substantial service to its licensed area within 10 years of initial license grant.	PC	80.49(a)(3)
Public Fixed	Applies only to licenses with station class of APC. Construct and place into operation within 12 months from date of grant.	MK	80.49(b)

Source: FCC.

Table 18. Buildout Requirements for Aviation Services, as of January 2014

Radio service/frequency band	Construction/coverage requirements	Radio service code	47 C.F.R.Section
Aeronautical advisory Unicom and Radionavigation Land Stations (excluding radionavigation land test stations)	Place into operation within 12 months after initial license grant.	AF, AR	87.45

Source: FCC.

Table 19. Buildout Requirements for Fixed-Microwave Services, as of January 2014

Radio service/frequency band	Construction/coverage requirements	Radio service code	47 C.F.R. Section
Fixed Microwave and Electronic Message Services (Site-Based)	Place in operation within 18 months of initial license grant.	CE, CF (except 39 GHz RSA), CT (Fixed locations only), MG, MW, PF, WA, WM, WR	101.63
Multichannel Video and Distribution and Date Services	Provide substantial service to its licensed area within 5 years of initial license grant and 10 years of initial license grant.	DV	101.1413
Local Multipoint Distribution Service	Provide substantial service to its licensed area within 10 years of initial license grant.	LD	101.1011
39 GHz Geographic Licenses	Provide substantial service to its licensed area within 10 years of initial license grant.	TN (39 GHz RSA)	101.17
24 GHz Geographic Licenses	Provide substantial service to its licensed area within 10 years of initial license grant.	TZ	101.527
Multiple Address Systems (Geographic)	Provide service to at least 1/5 of the population in its licensed area or substantial service within 5 years of initial license grant and make a showing of continued substantial service within 10 years of initial license grant.	MS	101.1325(b)

Source: FCC.

Table 20. Buildout Requirements for Broadcast Auxiliary Services, as of January 2014

Radio service/frequency band	Construction/coverage requirements	Radio service code	47 C.F.R.Section
Site-Based	Construct and place into operation within 18 months of initial license grant.	AB, AI, AS, TB, TI, TP, TS, TT	74.34(a)
Site-Based	Construct and place into operation within 12 months of initial license grant.	RP	74.34(b)

Source: FCC.

Table 21. Buildout Requirements for Personal Radio Service, as of January 2014

Radio service/frequency band	Construction/coverage requirements	Radio service code	47 C.F.R.Section
218-219 MHz (formerly Interactive Video and Data Service)	Provide substantial service within 10 years of license grant. File a report informing the Commission of the status of its system.	ZV	95.833(a)

Source: FCC.

Buildout Requirements for Services Licensed by the Media Bureau

Table 22. Buildout Requirements for Broadcast Services, as of January 2014

Radio service/frequency band	Construction/coverage requirements	47 C.F.R.Section
New TV, AM, FM, or International Broadcast; low power TV; TV translator; TV booster; FM translator; or FM booster station, or to make changes in such existing stations	Three years from the date of issuance of the original construction permit within which construction shall be completed and application for license filed.	73.3598
Low power FM station	Eighteen months from the date of issuance of the construction permit within which construction shall be completed and application for license filed.	73.3598

Source: FCC.

Buildout Requirements for Services Licensed by the International Bureau

Table 23. Buildout Requirements for Satellite Services, as of January 2014

Radio service/frequency band	Construction/coverage requirements	47 C.F.R. Section
New TV, AM, FM, or International Broadcast; low power TV; TV translator; TV booster; FM translator; or FM booster station, or to make changes in such existing stations	Three years from the date of issuance of the original construction permit within which construction shall be completed and application for license filed.	73.3598
Low power FM station	Eighteen months from the date of issuance of the construction permit within which construction shall be completed and application for license filed.	73.3598
Non-geostationary orbit satellite systems[a]	From the date the license is issued: 1 year: Enter into a binding non-contingent contract to construct the licensed satellite system. 2 years: Complete the critical design review of the licensed satellite system. 2 years, 6 months: Begin the construction of the first satellite in the licensed satellite system. 3 years, 6 months: Launch and operate the first satellite in the licensed satellite system. 6 years: Bring all the satellites in the licensed satellite system into operation.	25.164 25.137(d)[b]
Digital Audio Radio Services satellite systems	From the date the license is issued: 1 year: Complete contracting for construction of first space station or begin space station construction. 2 years: If applied for, complete contracting for construction of second space station or begin second space station construction.	25.144

Table 23. (Continued)

Radio service/frequency band	Construction/coverage requirements	47 C.F.R. Section
	4 years: In orbit operation of at least one space station. 6 years: Full operation of the satellite system.	
Direct Broadcast Satellite systems	From the date the license is issued: 1 year: Complete contracting for construction of the satellite station(s). 6 years: Satellite stations to be in operation.	25.148(b)

Source: FCC.

[a] Not to include Direct Broadcast and Digital Audio Radio Services satellite systems.

[b] Applies to non-U.S.-licensed satellite systems requesting access to the market in the United States.

Appendix III: Outcomes of Buildout Requirements for Selected Wireless Services

Tables 24 to 37 provide results from our analysis of FCC's ULS data for licenses for five selected wireless services. Among other things, these tables tabulate the number of licenses by auction, buildout requirement outcomes (e.g., whether a license met the buildout requirement), and extension request outcomes. Appendix I contains information on the scope and methodology of this analysis.

Table 24. Number of Licenses, by Wireless Service

Service	Number of licenses
Broadband PCS	4,212
220 MHz	1,231
39 GHz	2,173
Fixed microwave	10,561
Private land mobile radio, below 700 MHz, except 220 MHz (private land mobile radio)	10,278
Total	28,455

Source: GAO analysis of ULS data.

Table 25. Number of Market-Based Licenses, by Wireless Service and Auction

Service	Auction number and title	Auction dates	Number oflicenses
Broadband PCS	4: Broadband PCS A and B Block	12/05/1994-03/13/1995	760
	5: Broadband PCS C Block	12/18/1995-05/06/1996	640
	10: Broadband PCS C Block Reauction	07/03/1996-07/16/1996	15
	11: Broadband PCS D, E, & F Block	08/26/1996-01/14/1997	1,885
	22: C, D, E, and F Block Broadband PCS	03/23/1999-04/15/1999	483
	35: C and F Block Broadband PCS	12/12/2000-01/26/2001	171
	58: Broadband PCS	01/26/2005-02/15/2005	224
	71: Broadband PCS	05/16/2007-05/21/2007	34
		Service total	4,212
220 MHz	18: 220 MHz	09/15/1998-10/22/1998	859
	24: 220 MHz	06/08/1999-06/30/1999	292
	43: Multi-Radio Service	01/10/2002-01/17/2002	4
	72: 220 MHz	06/20/2007-06/26/2007	76
39 GHz	30: 39GHz	Service total	1,231
		04/12/2000-05/08/2000	2,173
		Service total	2,173
		Total	7,616

Source: GAO analysis of ULS data and FCC information.

Table 26. Number of Site-Based Licenses, by Wireless Service

Service	License grant dates	Number of licenses
Fixed microwave	07/01/2010 to 06/31/2011	10,561
Private land mobile radio	01/01/2011 to 12/31/2011	10,278
Total		20,839

Source: GAO analysis of ULS data and FCC information.

Table 27. Number of Licenses that Did and Did Not Have a Buildout Requirement, by Wireless Service

Service	Buildout requirement on or before 12/31/12	Percent	No buildout requirement	Percent	Buildout requirement after 12/31/12	Percent	Total
Broadband PCS	3,558	84.5%	650	15.4%	4	0.1%	4,212
220 MHz	963	78.2	212	17.2	56	4.5	1,231
39 GHz	1,672	76.9	501	23.1	0	0.0	2,173
Fixed microwave	10,478	99.2	22	0.2	61	0.6	10,561
Private land mobile radio	9,546	92.9	608	5.9	124	1.2	10,278
Total	26,217	92%	1,993	7%	245	1%	28,455

Source: GAO analysis of ULS data.

Notes: Licenses may have a buildout requirement after December 31, 2012, because, among other reasons, a licensee was granted a request for an extension to move the buildout deadline.

Percentages may total to more than 100 due to rounding.

Table 28. Buildout Requirement and Extension Request Outcomes for Select Broadband PCS Licenses

	Outcome	No request for extension	Granted but not dismissed request for extension	Dismissed but not granted request for extension	Total	Percentage of licenses with requirements
Met requirement	On time	2,972	62	41	**3,075**	86.4%
	Late	43	1	2	**46**[a]	1.3
Did not meet requirement	Terminated on deadline	40	10	7	**57**	1.6
	Not terminated on deadline	18	1	3	**22**[b]	0.6
	Terminated before deadline	11	0	0	**11**	0.3
	Canceled on or before deadline	292	1	54	**347**	9.8
Total for licenses with requirements		**3,376**	**75**	**107**	**3,558**	**100%**
Other	No buildout requirement	650	0	0	**650**	
	Deadline after 12/31/12	4	0	0	**4**	
Total for all licenses		**4,030**	**75**	**107**	**4,212**	

Source: GAO analysis of ULS data.

[a]Further research on these 46 licenses indicated that there were reasonable explanations for the buildout date being more than 15 days after the buildout deadline. For instance, FCC terminated 4 licenses and later returned them to active status in response to a petition for reconsideration. For another 21 licenses, FCC canceled and later reinstated the licenses. FCC "tolled" or suspended the buildout period to account for the time the licenses were not active, so the buildout date was more than 2 years after the buildout deadline.

[b] Of these 22 licenses, 6 licenses were canceled within 60 days after the buildout deadline before the automatic termination process was complete. However, the ULS information for the remaining licenses did not include information explaining why the licenses were not terminated on the buildout deadline.

Table 29. Number of Extension Requests for Select Broadband PCS Licenses

Number of extension requests	Total number of licenses	Percentage
0	4,030	95.7%
1	175	4.2
2	7	0.2
Total	4,212	100%

Source: GAO analysis of ULS data.

Note: Percentages may not total to 100 due to rounding.

Table 30. Buildout Requirement and Extension Request Outcomes for Select 220 MHz Phase II Licenses

	Outcome	No request for extension	Granted but not dismissed request for extension	Dismissed but not granted request for extension	Both Granted and dismissed request for extension	Total	Percentage oflicenses with requirements
Met requirement	On time	122	43	2	8	**175**	18.2%
	Late	6	0	0	0	**6**[a]	0.6
Did not meet requirement	Terminated on deadline	139	111	124	155	**529**	54.9
	Not terminated on deadline	105	29	0	0	**134**[b]	13.9
	Terminated before deadline	1	0	0	1	**2**	0.2
	Canceled on or before deadline	105	1	0	0	**106**	11.0
	Expired on or before deadline	3	1	4	3	**11**	1.1
Total for licenses with requirements		**481**	**185**	**130**	**167**	**963**	**100%**
Other	No buildout requirement	212	0	0	0	**212**	
	Deadline after 12/31/12	56	0	0	0	**56**	
Total for all licenses		**749**	**185**	**130**	**167**	**1,231**	

Source: GAO analysis of ULS data.

Note: Percentages may not total to 100 due to rounding.

[a]Further research on these 6 licenses indicated that FCC had terminated 3 licenses and later returned them to active status in response to petitions for reconsideration; the buildout dates entered in ULS were the dates of the petitions for reconsideration. For the other 3 licenses, the required notification was filed within 30 days of the buildout deadline with a waiver request to accept the late notification, in line with FCC policy.

[b] Of the 134 licenses that did not meet their buildout requirement and were not terminated, 110 licenses were canceled within 60 days after the buildout deadline, before the auto termination process was complete. Licensees for the remaining 24 licenses have submitted required notifications that are pending; FCC has yet to accept or dismiss the notifications. Licensees submitted the required notifications for 19 of these licenses before September 1, 2009.

Table 31. Number of Extension Requests for Select 220 MHz Phase II Licenses

Number of extension requests	Total number of licenses	Percentage
0	749	60.8%
1	315	25.6
2	141	11.5
3	21	1.7
4	5	0.4
Total	1,231	100%

Source: GAO analysis of ULS data.

Table 32. Buildout Requirement and Extension Request Outcomes for Select 39 GHz Licenses

	Outcome	No request for extension	Granted but not dismissed request for extension	Dismissed but not granted request for extension	Both granted and dismissed request for extension	Total	Percentage oflicenses with requirements
Met requirement	On time	0	837	0	0	**837**	50.1%
	Late	0	2	0	0	**2**[a]	0.1
Did not meet requirement	Terminated on deadline	0	144	0	345	**489**	29.2
	Not terminated on deadline	0	13	0	0	**13**[b]	0.8
	Canceled on or before deadline	0	308	0	0	**308**	18.4
	Expired on or before deadline	0	23	0	0	**23**	1.4
Total for licenses with requirements		**0**	**1,327**	**0**	**345**	**1,672**	**100%**
Other	No buildout requirement	496	5	0	0	**501**	
Total for all licenses		**496**	**1,332**	**0**	**345**	**2,173**	

Source: GAO analysis of ULS data.

[a]Further research on these 2 licenses provided no information on why the late-filed required notifications were accepted by FCC. For both licenses, the required notifications were filed 17 days after the buildout deadline.

[b]Licensees for all 13 licenses who did not meet their buildout requirement but whose licenses were not terminated have submitted a required notification that is pending; FCC has yet to accept or dismiss the notifications. Licensees submitted the required notifications for these 13 licenses between September 2011 and September 2012.

Table 33. Number of Extension Requests for Select 39 GHz Licenses

Number of extension requests	Total number of licenses	Percentage
0	496	22.8%
1	1,321	60.8
2	356	16.4
Total	2,173	100%

Source: GAO analysis of ULS data.

Table 34. Buildout Requirement and Extension Request Outcomes for Select Fixed-Microwave Licenses

	Outcome	**No request for extension**	**Granted but not dismissed request for extension**	**Dismissed but not granted request for extension**	**Total**	**Percentage oflicenses with requirements**
Met	On time	6,195	10	4	**6,209**	59.3%
	Late	90	0	0	**90**[a]	0.9
Both met and did not meet	Met for some and did not meet for other frequencies	1,215	0	1	**1,216**	11.6
Did not meet	All frequencies terminated	2,172	0	7	**2,179**	20.8
	Some frequencies terminated/some without requirement	62	0	4	**66**	0.6
	Not terminated on deadline	106	0	0	**106**[b]	1.0
	Terminated before deadline	12	0	0	**12**	0.1
	Canceled on or before deadline	600	0	0	**600**	5.7

	Outcome	No request for extension	Granted but not dismissed request for extension	Dismissed but not granted request for extension	Total	Percentage oflicenses with requirements
Total for licenses with requirements		**10,452**	**10**	**16**	**10,478**	**100%**
Other	No buildout requirement	22	0	0	**22**	
	Deadline after 12/31/12	47	14	0	**61**	
Total for all licenses		**10,521**	**24**	**16**	**10,561**	

Source: GAO analysis of ULS data.

Note: Includes licenses granted between July 1, 2010, and June 30, 2011.

[a]Further research on a sample of 9 of these licenses indicated that FCC approved a petition for reconsideration for each of the 9 met-late licenses.

[b]We examined the number of days between the cancellation dates and the buildout deadlines for the licenses that did not meet their buildout requirement but were not terminated; since these licenses can have more than one buildout deadline, we constructed a proxy buildout deadline 584 days (i.e., 18 months) from a license grant date. We found that all the licenses had a cancellation date within 60 days of the proxy buildout deadline, meaning that the licensee canceled the license before the automatic termination process was completed.

Table 35. Number of Extension Requests for Select Fixed-Microwave Licenses

Number of requests for a license	Total number of licenses	Percentage
0	10,521	99.6%
1	39	0.4
2	1	0.0
Total	10,561	100%

Source: GAO analysis of ULS data.

Table 36. Buildout Requirement and Extension Request Outcomes for Select Industrial/Business Private Land-Mobile-Radio Licenses

	Outcome	**No request for extension**	**Granted but not dismissed request for extension**	**Dismissed but not granted request for extension**	**Both granted and dismissed request for extension**	**Total**	**Percentage ofall licenses with requirements**
Met	On time	8,862	17	1	1	**8,881**	93.0%
	Late	261	0	0	0	**261**[a]	2.7
Both met and did not meet	Met for some and did not meet for other frequencies	13	0	0	0	**13**	0.1
Did not meet	All frequencies terminated	324	0	2	0	**326**	3.4
	Some frequencies terminated/some without requirement	17	0	0	0	**17**	0.2
	Not terminated on deadline	9	0	0	0	**9**[b]	0.1
	Terminated before deadline	12	0	0	0	**12**	0.1
	Canceled on or before deadline	27	0	0	0	**27**	0.3

	Outcome	No request for extension	Granted but not dismissed request for extension	Dismissed but not granted request for extension	Both granted and dismissed request for extension	Total	Percentage ofall licenses with requirements
Total for licenses with requirements		**9,525**	**17**	**3**	**1**	**9,546**	**100%**
Other	No buildout requirement	608	0	0	0	**608**	
	Deadline after 12/31/12	37	85	0	2	**124**	
Total for all licenses		**10,170**	**102**	**3**	**3**	**10,278**	

Source: GAO analysis of ULS data.

Note: Includes licenses granted between January 1, 2011, and December 31, 2011.

[a]Further research on a sample of 27 licenses that met their buildout requirements late indicated that FCC approved a petition for reconsideration or a waiver to file late for 25 licenses. The online ULS records for the remaining 2 licenses did not contain information explaining why the buildout date was more than 15 days after the buildout deadline; the required notifications for these 2 licenses were 17 and 26 days late.

[b] We conducted further research on the 9 licenses that did not meet their buildout requirement but were not terminated, including examining the number of days between the cancellation dates and the buildout deadlines. Since these licenses can have more than one buildout deadline, we constructed a proxy buildout deadline 365 days from each license grant date. Six of the licenses were canceled after their buildout deadlines, all within 60 days of the proxy buildout deadline, meaning that the licensee canceled the license before the automatic termination process was complete. The other 3 licenses were active as of September 1, 2013, but all 3 were in termination-pending mode.

Table 37. Number of Extension Requests for Select Industrial/Business Private Land-Mobile-Radio Licenses

Number of extension requests	Total number of licenses	Percentage
0	10,170	98.9%
1	62	0.6
2	44	0.4
3	2	0.0
Total	10,278	100%

Source: GAO analysis of ULS data.

Note: Percentages may not total to 100 due to rounding.

End Notes

[1] For many mobile radio systems, the 300 megahertz (MHz) to 3 gigahertz (GHz) spectrum range is the portion of the spectrum where scarcity concerns are the greatest. While other, higher-frequency spectrum is available as green field or underutilized spectrum, these bands are less favorable for traditional wireless networks.

[2] The Department of Commerce's National Telecommunications and Information Administration (NTIA) is responsible for managing the federal government's use of the radio frequency spectrum.

[3] A band is a range of frequencies in the spectrum that are used or set aside for a specific purpose (such as the 1,710–1,755 MHz band that FCC allocated for fixed and mobile services and designated for Advanced Wireless Services to support Third Generation (3G) mobile broadband and advanced-wireless services).

[4] 47 C.F.R. § 1.946(c).

[5] Wireless services include cellular telephone, paging, personal communications services, and other commercial and private radio services. FCC also sets buildout requirements for satellite, broadcast, and public safety communications spectrum licenses, which we did not include in the scope of our review.

[6] ULS operates as a single, seamless licensing system for all wireless services to process licenses in a uniform manner. ULS also serves as the central, consolidated repository for application and license data for wireless services.

[7] Due to the high volume of licenses in the fixed microwave and industrial/business private land mobile radio services, we limited our analysis to new licenses with buildout requirements during calendar year 2012.

[8] Communications Act of June 19, 1934, ch. 652, as amended, codified at 47 U.S.C. ch. 5, subch. II (the Communications Act).

[9] National Telecommunications and Information Administration Organization Act, Pub. L. No. 102-538, title I, 106 Stat. 3533, codified as amended at 47 U.S.C. ch. 8.

[10] 5 U.S.C. § 553.

[11] The public dockets for FCC are maintained on the Electronic Comment Filing System, which is accessible on the Internet at http://apps.fcc.gov/ecfs.

[12] Since 1994, when FCC first implemented its auction authority, spectrum auctions have generated nearly $52 billion in net offsetting receipts for the U.S. Treasury.

[13] Congress authorized FCC to impose and collect application processing fees. 47 U.S.C. § 158. These application fees are intended to cover FCC's costs related to licensing communication service providers. For example, as of August 23, 2013, the new license fee for a private point-to-point microwave license is $270.

[14] FCC has not set buildout requirements for only a few wireless services including personal radio services like citizens band (CB) radio and family radio services, which both use two-way communications similar to walkie-talkies. An individual license is not required to use many personal radio services; rather, FCC certifies devices that may be used for these services, generally through the rulemaking process (commonly termed "licensing by rule"). Anyone who is eligible is thus authorized to purchase and operate technically compliant and FCC-approved devices in accordance with the operational requirements of the rules.

[15] Different frequency radio signals propagate differently. Signals at lower frequencies, for instance, can travel through buildings and other objects better than signals at higher frequencies. Different propagation characteristics are better suited for different services, and can thus influence the licensee's infrastructure needs and costs.

[16] "Substantial service" is defined in regulation as "service which is sound, favorable, and substantially above a level of mediocre service which just might minimally warrant a renewal." 47 C.F.R. § 22.940(a)(1)(i).

[17] There is an intermediate step for terminating site-based licenses. Each frequency authorized by a license is identified by a location number. So, as noted, if a buildout requirement for a frequency is not met, then the frequency is terminated. Next, if all of the frequencies at a particular location are terminated, the location is terminated. Finally, if all of the locations for a license are terminated, then the license is terminated. Also, as noted in the next section, FCC places licenses in a termination-pending mode for 30 days before terminating a license.

[18] Licensees must notify FCC within 15 days of the buildout deadline. 47 C.F.R. § 1.946(d). FCC officials noted that the Commission's rules provide that filings due on a weekend, holiday, or day that the Commission closes early and does not reopen may be timely filed on the next business day; as a result, some required notifications dated more than 15 days after a license's buildout deadline may be timely.

[19] The ULS auto termination process was implemented on February 1, 2006, for licenses with buildout deadlines that fell on or after that date.

[20] A license is made available for such reassignment unless FCC rules provide that the spectrum covered by the license would automatically revert to another license holder.

[21] FCC may also take enforcement action in response to complaints filed by third parties or based on information obtained through other avenues.

[22] 47 C.F.R. § 1.946(e).

[23] Specifically, FCC extended the buildout deadline for (1) all 220 MHz phase II economic area, regional, and nationwide licenses that timely sought an extension of the applicable construction deadline, if the deadline had expired as of the date of the release of the FCC order granting the extension; (2) all 220 MHz phase II economic area, regional, and nationwide licenses with 5-year construction deadlines falling after the release date of the order; and (3) the phase II 220 MHz licenses covered by a notification of substantial service filed by one entity. *Dismissal of 220 MHz Service Phase II Construction Notifications,* 19 FCC Rcd. 18,095 (2004); *Request of Warren C. Havens for Waiver or Extension of the Five Year Construction Requirement for 220 MHz Service,* 19 FCC Rcd. 12,994 (2004).

[24] These 106 licenses make up the remaining 1 percent of fixed-microwave licenses in the other-outcomes category in Table 4. For private land mobile radio, the 9 remaining other-outcome licenses include 6 licenses that were canceled during the auto termination process and 3 licenses that were in termination-pending mode as of September 1, 2013.

[25] The remaining 6 licenses in the other-outcomes category were canceled during the auto termination process.

[26] For example, a license may be automatically terminated if the licensee has failed to meet a payment condition for a license obtained through the auction process.

[27] A licensee may request more than one extension of a buildout deadline, thus a licensee could have requests both granted and dismissed.

[28] See 47 C.F.R. §§ 1.925, 1.946.

[29] FCC officials told us that buildout requirements are one tool in a framework used to meet its broad goal of promoting the efficient use of spectrum. Beyond this more general goal, we reviewed statutes, FCC budget and performance documents, and FCC reports and orders from recent rulemaking proceedings to identify four commonly cited goals for buildout requirements for wireless-service licenses.

[30] Of the remaining experts, two said that the requirements were effective, one replied don't know, and two did not provide an opinion for this goal.

[31] Specifically, three experts said effective, one said neither, and three said ineffective, while two experts did not provide an opinion for the goal.

[32] Stakeholders that identified changes to current buildout requirements included 9 of 10 licensees, 6 of 9 associations, and 7 of 9 experts.

[33] Greater detail could include setting safe harbors for particular wireless services with substantial service requirements. As noted earlier, FCC has established safe harbors for some services; safe harbors illustrate specific ways that a licensee could demonstrate substantial service, such as constructing a certain number of point-to-point links or serving populations that are outside areas served by other licensees.

[34] We reviewed FCC dockets 07-293 *In the Matter of Establishment of Rules and Policies for the Digital Audio Radio Satellite Service in the 2310-2360 MHz Frequency Band,* and 12-268 *In the Matter of Expanding the Economic and Innovation Opportunities of Spectrum through Incentive Auctions.* FCC requested comments on the proposed buildout requirements in the draft rules. The comment periods for these proceedings closed in 2010 and 2013, respectively.

[35] See also GAO, *Telecommunications: Enhanced Data Collection Could Help FCC Better Monitor Competition in the Wireless Industry*, GAO-10-779 (Washington, D.C.: July 27, 2010).

[36] See also GAO, *Spectrum Management: Incentives, Opportunities, and Testing Needed to Enhance Spectrum Sharing*, GAO-13-7 (Washington, D.C.: Nov. 14, 2012).

[37] See also GAO-13-7.

[38] In 2011, the FCC set a budget of $4.5 billion annually over six years (2011-2016) for the high-cost program within the Universal Service Fund and, among other reforms, established support for mobile voice and broadband. GAO, *Telecommunications: FCC Has Reformed the High-Cost Program, but Oversight and Management Could be Improved,* GAO-12-738 (Washington, D.C.: July 25, 2012).

[39] See GAO-13-7. In this report, FCC officials said that the Commission has proposed spectrum usage fees at various times including in its congressional budget submission, but FCC has not received legislative authority to implement such a program.

End Notes for Appendix I

[1] Wireless services include cellular telephone, paging, personal communications services, and other commercial and private radio services. FCC also sets buildout requirements for satellite, broadcast, and public-safety communications spectrum licenses that are administered by other bureaus, which we did not study in detail for this review.

[2] FCC uses radio service codes to categorize licenses by wireless service in ULS, which we used to identify licenses in our selected wireless services. For the three market-based services, broadband PCS licenses are coded CW, 220 MHz phase II licenses are coded QA, and 39 GHz licenses are coded TN. For the two site-based services, multiple radio service codes corresponded to each selected service. We excluded radio service codes for public safety licenses or licenses that have a strong or dominant federal use, such as commercial aviation. As a result, only industrial/business private land-mobile-radio licenses coded IG and YG and fixed-microwave licenses coded CF and MG were included for these two services.

[3] For the selected wireless services, we excluded licenses for special temporary authority (STA). According to FCC, STA licenses are granted for short periods of time, including emergency situations, so the licenses are not subject to buildout requirements.

[4] All broadband PCS licenses have a 5-year buildout requirement either as a first requirement (for 30 MHz licenses) or as the only requirement (for 10 and 15 MHz licenses). We excluded broadband PCS licenses awarded at auction 78 in 2008 from the analysis since no buildout requirements would come due until 2013.

[5] FCC updates ULS database downloads for specific wireless radio services on a weekly basis. The downloads are available at http://wireless.fcc.gov/uls/index.htm?job=transaction&page=weekly.

[6] Licensees must notify FCC within 15 days of the buildout deadline. 47 C.F.R. § 1.946(d). We categorized a license as meeting the buildout requirement "on time" if its buildout date was within 15 days of the buildout deadline. FCC officials noted that the Commission's rules provide that filings due on a weekend, holiday, or day that the Commission closes early and does not reopen may be timely filed on the next business day; as a result, some required notifications dated more than 15 days after a license's buildout deadline may be timely.

[7] We considered experts interviewed for GAO, *Commercial Spectrum: Plans and Actions to Meet Future Needs, Including Continued Use of Auctions,* GAO-12-118 (Washington, D.C.: Nov. 23, 2011), *Spectrum Management: Incentives, Opportunities, and Testing Needed to Enhance Spectrum Sharing,* GAO-13-7 (Washington, D.C.: Nov. 14, 2012), and *Spectrum Management: FCC's Licensing Approach in the 11, 18, and 23 Gigahertz Bands Currently Supports Spectrum Availability and Efficiency*, GAO-13-78R (Washington, D.C.: Nov. 20, 2012).

[8] Federal Communication Commission Requests Comment on Revision of Performance Requirements for 2.3 GHz Wireless Communications Service, FCC document 10-46 (2010).

[9] *In the Matter of Expanding the Economic and Innovation Opportunities of Spectrum Through Incentive Auctions*, FCC document 12-118 (2012). Due to the high number of filings in this docket, we did not review filings made by individuals.

INDEX

#

A

B

C

D

E

F

G

H

I

J

K

L

M

N

O

P

Q

R

S